Successfi forestry

A Guide to Private Forest Management

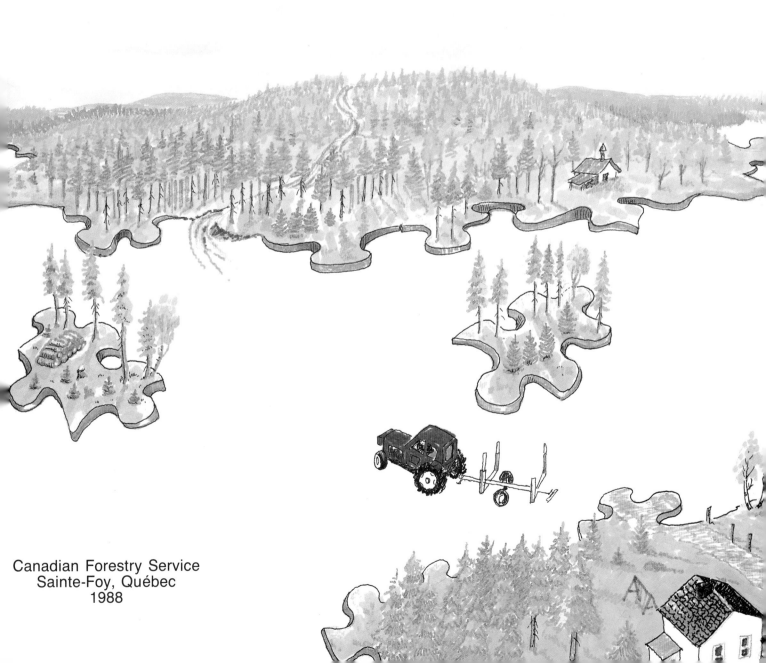

Canadian Forestry Service
Sainte-Foy, Québec
1988

Copies of this publication may be purchased from the:

Canadian Government Publishing Centre
Supply and Services Canada
Ottawa, Ontario
Canada
K1A 0S9

Cette publication est également disponible en français sous le titre
«Réussir ma forêt — Guide d'aménagement des forêts privées»

FOREWORD

A Guide to Private Forest Management has been prepared for woodlot owners who wish to practice forest management on their woodlots.

The Guide, which contains twelve sections, covers a variety of topics and is intended to help woodlot owners master basic forestry and silvicultural techniques so that they can enhance the value of their property by applying fundamental principles of forest management.

This Guide has been designed for easy consultation and to take into account your individual work pace and your most pressing needs. The liberally illustrated text will familiarize you with various silvicultural treatments.

Obviously, we do not claim that this Guide covers every aspect of forest management. Nevertheless, we do hope that it will provide you with adequate information and encourage you to manage your woodlot properly. Last but not least, it is always a pleasure for our forestry advisers to be of assistance.

On behalf of the Canadian Forestry Service, I hope that this new tool will help put you on the right track to "Successful Forestry".

Yvan Hardy, Professional Forester, Ph. D.
Director General
Quebec Region

TABLE OF CONTENTS

Management is a very general term that is used in many fields. This section explains what woodlot management is and why it is important. It also describes the various steps involved in woodlot management.

WOODLOT MANAGEMENT

Woodlot owners very often want their forests to be a source of additional income. There are two ways to get that income: harvest the wood for a quick profit or plan the harvest to produce a steady, indeed steadily increasing, income over the years while ensuring the woodlot's future. Proper management makes the second choice possible. Furthermore, good management is consistent with the objectives of maintaining water quality and improving wildlife habitat.

What is woodlot management?

Woodlot management involves a series of well-considered operations designed to improve the woodlot's productivity and thus maximize the benefits. Unlike farmers, woodlot owners do not always have a clear idea of what they want to produce. A farmer, for example, sets aside some areas for grazing and others for growing grain.

In order to manage your woodlot, you must first establish production objectives and then determine how to meet them, taking into account your woodlot's present state and potential.

Before you can manage your woodlot, you must first ask yourself three questions:

> ● **What do I want to do?**
> **(Personal objectives)**

First, decide what you want to produce. Will it be pulpwood, lumber, or other products? The woodlot can be used for purposes other than producing wood products — for example, recreational activities.

> ● **What can I do?**
> **(Woodlot potential)**

What tree species are suited to the soil and climate? Are the trees of high enough quality to produce lumber? Is there a market for the product? Do you have the right tools or will you have to get them?

> ● **What do I have to do?**
> **(Silvicultural practices)**

Once you have established your objectives, you must determine how to meet them. Decide what steps to take to reach your goals and when and in what order you will take these steps.

Why manage your woodlot?

Most woodlot owners manage their woodlots in order to:

- *obtain a worthwhile income from the woodlot;*

- *increase productivity;*

- *enhance product quality; and*

- *preserve and improve the forest for themselves and for future generations.*

When you manage a woodlot to produce wood products, you must also promote conservation and improve the habitat for both wildlife and fish.

In this way, you also contribute to the community's welfare, because forests provide many socio-economic benefits. It must not be forgotten that a private forest is of enormous potential to industry and the community. Private forests are close to sources of labor and to industries. They often occupy very productive land.

Think of the forest as money in the bank. Each year, a certain amount of wood grows. This annual growth, which can be harvested in the short, medium, or long term, represents the interest. Management increases annual growth, which means capital and interest income increase. So you get more benefits from the woodlot.

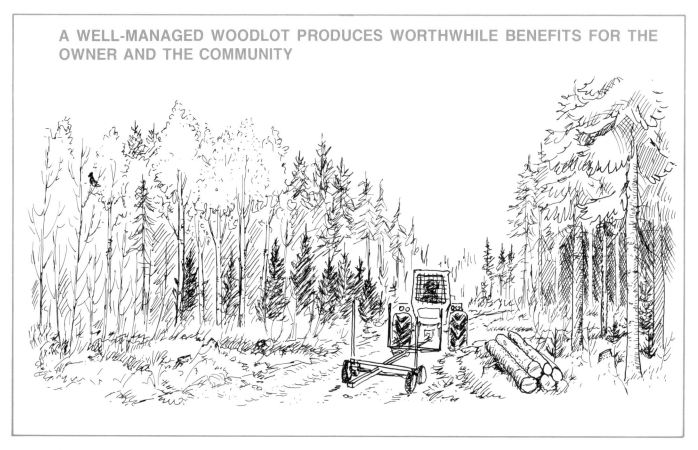

A WELL-MANAGED WOODLOT PRODUCES WORTHWHILE BENEFITS FOR THE OWNER AND THE COMMUNITY

How to manage a woodlot

There are various steps involved in managing a woodlot. These steps are described in the management plan. The plan is a guide to help you:

1. *learn more about your woodlot;*

2. *decide what steps to take;*

3. *plan effectively and reduce the risk of errors, which are sometimes difficult to correct; and*

4. *keep an inventory of the work done.*

In order to develop a management plan for your woodlot, you, with the help of your forestry adviser, must take the following steps:

- **Knowing your woodlot**

  To learn about your woodlot, you must draw up an inventory of the tree species, volumes, areas, and ages of the stands. In other words, you must evaluate the different characteristics of the woodlot in order to decide what steps to take.

- **Making choices**

  Once you are familiar with your woodlot, you can establish your management objectives. Now is the time to decide what you want to produce, be it lumber, veneer wood, maple products, or something else.

In managing your woodlot, you should try to obtain a variety of products from a variety of tree species. Diversification increases your chances of having a stable income and reduces the risk of damage caused by insects, diseases, and other forest enemies. This diversification must take into account the potential of the area and the region.

KNOWING YOUR WOODLOT IS AN ESSENTIAL PART OF GOOD MANAGEMENT

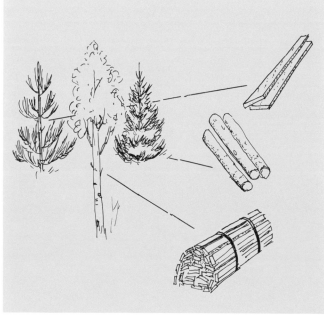

- **Planning**

  Now that you know what you want to produce, you must decide how you will produce it (for example, reforestation, thinning, or conversion).

  You must also decide when and where to do the work.

Woodlot managment is not an easy task. It requires technical knowledge, which you can get from your forestry adviser, training courses, or documentation such as this guide. Nevertheless, the final choice is yours. You are the one who will decide, on the basis of the advice you receive, why, how, and when to manage your woodlot.

WORK TOOLS AND METHODS

Forest work requires some familiarity with appropriate tools and methods. Safety should be your first consideration. This section suggests techniques to make your work both easier and safer.

WORK TOOLS AND METHODS

Following a few rules when you use forestry tools or equipment will make your work easier. Always use the proper tools for the job. Plan what you are going to do. And always use safe, proven methods.

- safety gloves or mittens;

SAFETY EQUIPMENT

Forestry tools can be dangerous. Always use them carefully and practice safe work methods.

Safety equipment reduces the risk of injury. Always wear:

- leg protectors (safety leggings or pants).

Your chain saw should have an emergency brake and an anti-vibration handle.

Finally, you should always have a first-aid kit and a dry chemical extinguisher at hand.

- an approved safety helmet;

- a safety visor or goggles;

- ear protectors to reduce noise;

Anti-vibration handle

- safety boots;

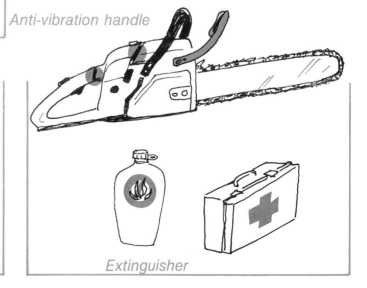

Extinguisher

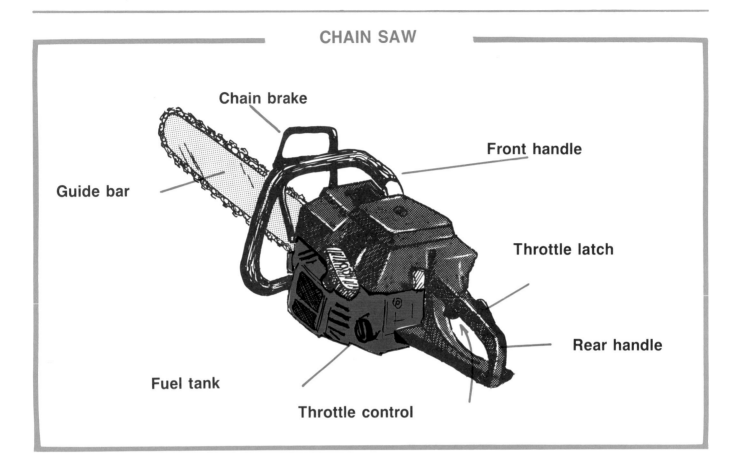

CHAIN SAW

Chain brake

Front handle

Guide bar

Throttle latch

Rear handle

Fuel tank

Throttle control

General maintenance

A chain saw is subject to considerable stress when in use. Proper maintenance is therefore essential.

The most important thing is to keep the chain sharp. Sharpen it as often as necessary; sometimes this will be three or four times a day. Keep the chain sharp and taut while working. Make sure that the braking mechanism is working properly, the oil pan is always full, and the chain is being properly lubricated.

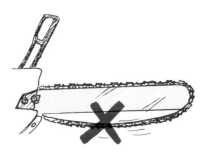

EVERY DAY

- Clean the air filter with soapy water or clean gasoline;

- Turn the guide bar over to get even wear on both sides;

- Clean the guide bar oil hole and slot;

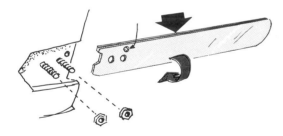

- Check that all parts of the braking system are working properly.

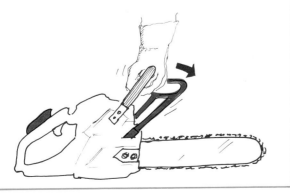

EVERY WEEK

- Check the spark plug and the starter mechanism;

Starter housing   **Spring**

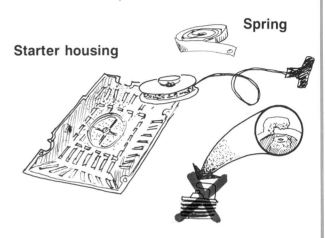

- Check that the clutch, guide bar, chain, and oil filter are in good shape;

Oil filter

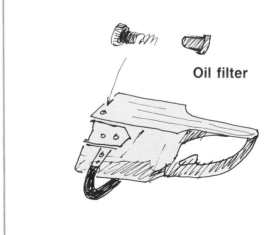

Follow all the instructions in the manufacturer's guide for the chain saw.

SHARPENING THE CHAIN

To work safely and efficiently, ensure that the chain is sharp. Poor maintenance can lead to accidents or breaks.

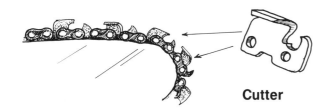

Cutter

The method of sharpening depends on the kind of edge the cutter has; the edge can be straight or rounded.

ROUNDED	STRAIGHT

35°

30°

90°

10°

Always use a file designed especially for sharpening chain saws. The size of file depends on the saw pitch.

Saw pitch = this distance divided by 2

Saw pitch	File diameter
1/4''	5/32''
0.325''	3/16''
3/8''	7/32''
0.404''	7/32''

The depth guide must be lower than the cutting edge. This is what determines the depth of the cut.

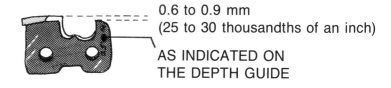

0.6 to 0.9 mm
(25 to 30 thousandths of an inch)

AS INDICATED ON
THE DEPTH GUIDE

Adjust the height of the guide every third time you sharpen the saw. Always use a depth gauge to correct height of the depth guide.

Depth gauge

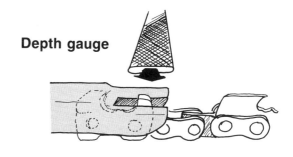

If the depth guide is too low or is not rounded, it increases the risk of kickback.

Remember that only light pressure with the file is needed to sharpen the edges. Always file in a forward direction.

FELLING TECHNIQUE

The direction in which a tree falls is affected by the tree's lean, the shape and arrangement of branches, the wind direction and, sometimes, the weight of snow. Evaluate each of these factors to decide in which direction you should fell a tree.

Look out for dead branches

Once you choose a direction, clear the space around the tree and cut off the lower branches.

Then make the undercut, starting with the upper part of the cut. The undercut should have a minimum angle of 45° and should extend at least 1/3 of the way through the trunk.

Next, make the backcut 2 to 4 cm above the back of the undercut.

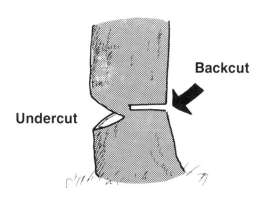

Backcut

Undercut

The holding wood between the back of the undercut and the backcut acts as a hinge and assists in felling the tree. It is the line of undercut, not that of the backcut, that determines which way the tree falls.

You cannot compensate for a bad undercut by leaving a thicker hinge of holding wood on one side.

The hinge should be at least 3 cm thick. If the tree has rot, compression wood, or spiral grain, make the hinge thicker.

Hinge

BE CAREFUL

The most dangerous moment is when the tree starts to fall. Stop cutting, pull the saw out and move away on the diagonal within the clearance zone.

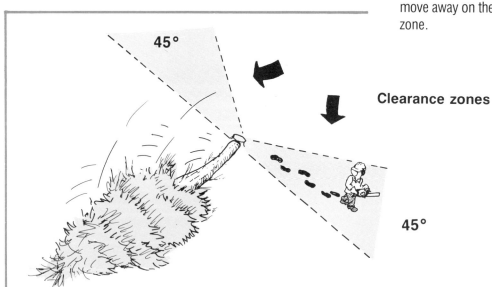

45°

Clearance zones

45°

14

LIMBING TECHNIQUE

Of all the operations performed with a chain saw, limbing has the most hidden dangers.

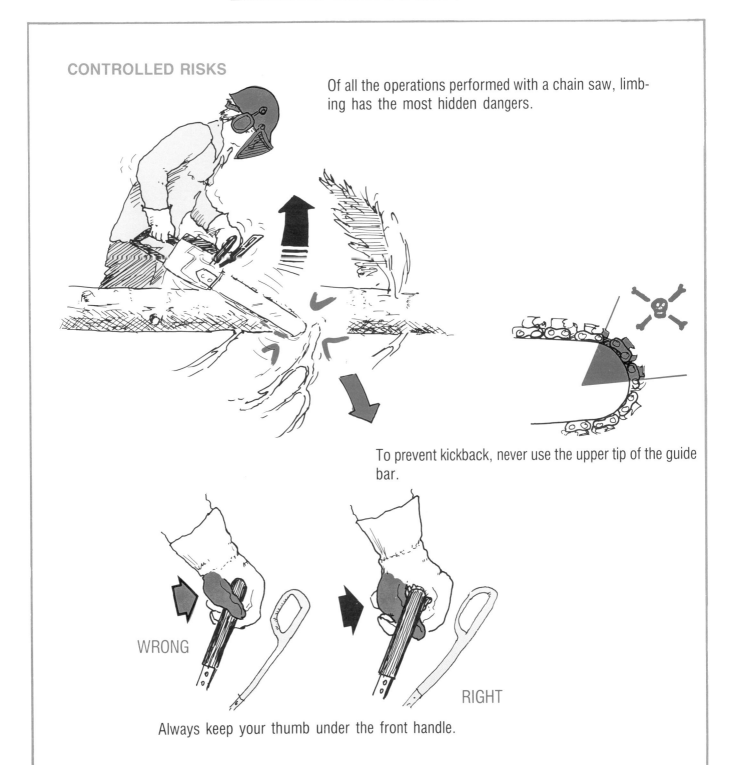

To prevent kickback, never use the upper tip of the guide bar.

WRONG

RIGHT

Always keep your thumb under the front handle.

Basic rules

Co-ordinate your movements with the saw's to make the work easier, safer, and more productive.

Hold the saw at a height between your hips and knees.

Hold the saw close to you. The sawdust will be thrown off to the side. You will be able to control the saw more easily and you will find it more comfortable.

Rest the weight of the saw on the trunk. Use the saw as a lever.

Keep the guide bar in contact with the trunk at all times.

- **Work at a comfortable height**

- **Use the saw as a lever**

- **Rest the saw on the trunk**

- **Keep the guide bar in contact with the trunk at all times**

BUCKING TECHNIQUE

Safety first

Assume a good working position. Facing the trunk, stand to the side of the section you are cutting. Have an escape path ready in case the trunk tips over.

Check the tension in the trunk and work inside the curves.

If the trunk curves downward, make your first cut on top and then cut from below. If the trunk curves upward, do the opposite. This will prevent the guide bar from jamming and breaking.

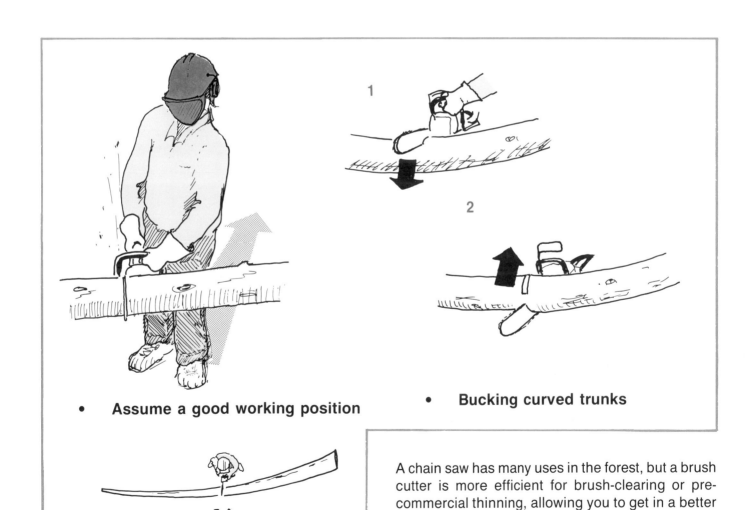

- **Assume a good working position**

1

2

- **Bucking curved trunks**

- **Work inside the curves**

A chain saw has many uses in the forest, but a brush cutter is more efficient for brush-clearing or pre-commercial thinning, allowing you to get in a better working position. With the proper technique and a little practice, you will be capable of directional felling.

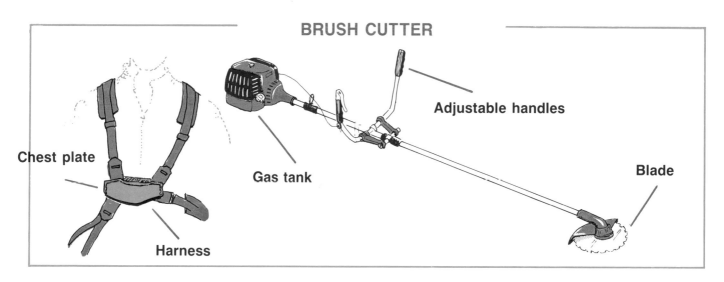

BRUSH CUTTER

Chest plate

Gas tank

Harness

Adjustable handles

Blade

Adjusting the tool

Adjust the harness and brush cutter properly so that you can work with more control and less effort.

* Position the chest plate in the center of your chest;

* Clip the brush cutter onto the harness about 10 cm (4 in.) below the hip;

Sharpening the blade

Good performance depends on a sharp blade. For the blade to turn smoothly and cut properly, the teeth must also be of uniform length.

You can use either a flat file with rounded edges or a round file.

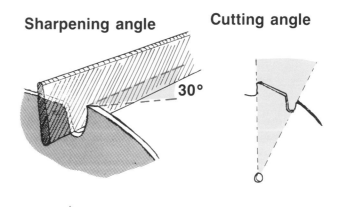

40 cm

Sharpening angle

Cutting angle

30°

* Keep the gas tank half full and use it to balance the weight of the brush cutter. Keep the blade in front of you and about 40 cm (15 in.) from the ground;

* Adjust the handles so that your arms are slightly bent.

The sharpening angle should be 30° and the cutting angle should be as close as possible to the blade's central axis. This will provide superior performance and smoother operation.

Finally, to maintain the resistance of the teeth, gradually reduce their length.

Sawing and felling technique

Directional felling is possible with a brush cutter. This makes movement easier and speeds up work.

Think of the blade as a watch face.

Cutting with the left half of the blade tends to pull the blade into the trunk; cutting with the right half, however, tends to pull it out of the trunk.

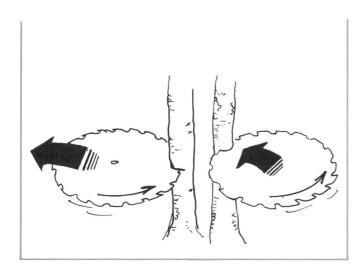

Never use the section of the blade between twelve and two o'clock. This can cause kickback or jamming.

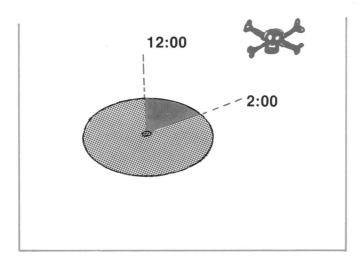

1. *To fell forward and to the right*

- tilt the blade to the left;

- use the section of the blade at eight o'clock; and

- pull it toward you.

2. *To fell backward and to the right:*

- tilt the blade to the left;

- use the section of the blade at three o'clock; and

- push it to the right.

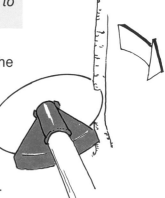

3. *To fell backward and to the left:*

- tilt the blade to the right;

- use the section of the blade at two-thirty; and

- push it forward.

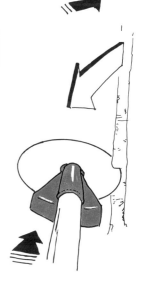

OTHER TOOLS

There are other tools that are useful in forestry work. These too must be used safely.

Ax, machete, and swedish brush ax

These tools are used to cut branches or clear brush over small areas. Always wear safety leggings and gloves when working with them.

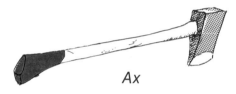

Ax

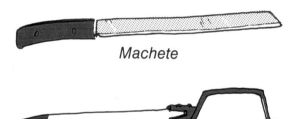

Machete

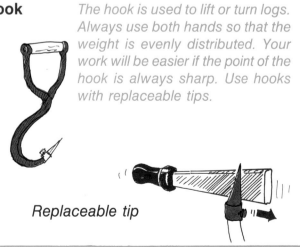

Swedish brush ax

Hook

The hook is used to lift or turn logs. Always use both hands so that the weight is evenly distributed. Your work will be easier if the point of the hook is always sharp. Use hooks with replaceable tips.

Replaceable tip

Wedge, felling lever

These tools are used to help fell trees.

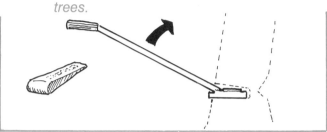

Pruning shears and saw

These are used to prune trees. Always wear gloves when working with them.

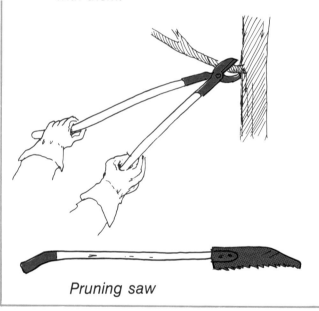

Pruning saw

Whenever you use any of these tools, wear a safety helmet, visor or goggles, and safety boots. Always carry a first-aid kit.

FARM TRACTOR

A farm tractor can be used to harvest wood. However, the tractor must first be modified, since it was not designed for forestry work.

The tractor should have:

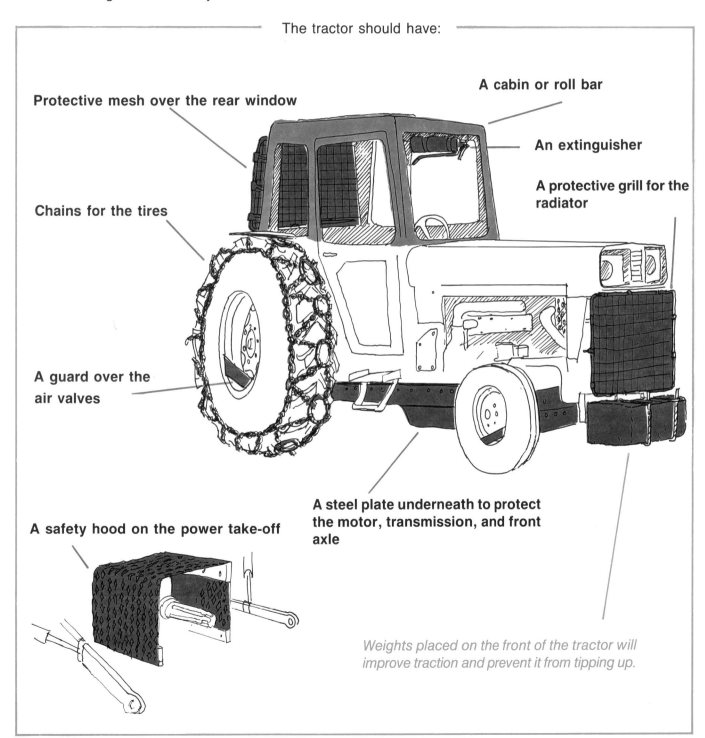

Protective mesh over the rear window

A cabin or roll bar

An extinguisher

A protective grill for the radiator

Chains for the tires

A guard over the air valves

A safety hood on the power take-off

A steel plate underneath to protect the motor, transmission, and front axle

Weights placed on the front of the tractor will improve traction and prevent it from tipping up.

CONCLUSION

We have shown you the basic operating principles for various tools. There is other material available that can provide more complete information. Do not forget that maintaining your tools and using the right tool in the right way will make your job easier. Finally, always keep in mind that safety and performance go hand in hand.

SITE PREPARATION

One important procedure which must be carried out before reforestation work can begin is site preparation. This is a silvicultural practice intended to ensure optimum growing conditions for your seedlings. It also makes reforestation easier.

SITE PREPARATION

You have probably already seen clear-cut sites. They are usually covered with stumps and piles of branches, as well as a great deal of grass and brush (including such species as raspberry, mountain maple, and pin cherry).

Such conditions sometimes make reforestation impossible and often threaten your seedlings' chances of survival. They also tend to hinder the forest workers.

Use of an appropriate silvicultural practice — site preparation — makes it possible to reach the following three objectives:

- *Reduce the amount of logging debris (slash);*

- *Improve soil conditions to encourage rooting and growth of seedlings;*

- *Temporarily eliminate undesirable vegetation;*

The various procedures involved in site preparation will help you accomplish these objectives.

Regardless of the method used, the site must be prepared in such a way that a seedling can be planted every 2 m (6.5 ft.), producing a final density of 2 500 seedlings per hectare (1 000 seedlings per acre).

PRINCIPLES OF SITE PREPARATION

Clearing away logging debris (slash)

Clearing away logging debris (slash) consists of removing branches and other material left on the ground after cutting, and collecting this material in piles called windrows. This process ensures that the ground is free of debris and thus makes reforestation easier. Otherwise, the piles of debris would impede progress and make it difficult to select planting sites (microsites).

Soil scarification

Scarification loosens the soil by mixing the mineral layer with the humus to a depth of 5 to 15 cm (2 to 6 in.)

Mixing the soil in this way improves its moisture absorption and retention capacity, making water more readily available for the seedlings. Moreover, in places where the humus layer is very deep, scarification ensures that the seedling roots come into contact with the minerals contained in the soil. Lastly, this process increases the temperature of the soil slightly, thereby encouraging root development.

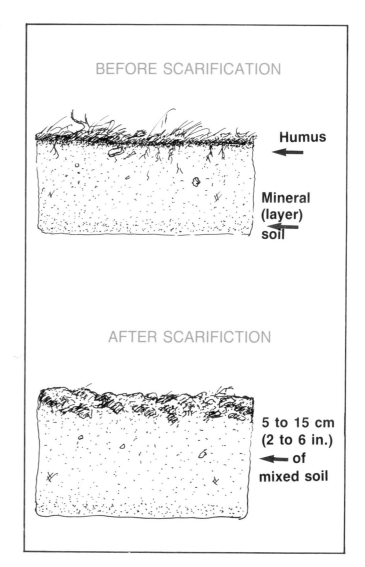

BEFORE SCARIFICATION

Humus

Mineral (layer) soil

AFTER SCARIFICTION

5 to 15 cm (2 to 6 in.) of mixed soil

Vegetation control

Seedlings need three elements to survive:

- light

- water

- nutrients.

Your seedlings must compete with undesirable vegetation for these elements. The competition is unequal, however, since grass and brush usually grow faster than seedlings. The resulting difficult conditions often cause the seedlings to die.

It is therefore important to control the development of competing vegetation before attempting to reforest the site.

The various techniques used rarely result in total eradication of competing vegetation; rather, they control its growth, making it easier to plant the seedlings and encouraging their survival.

SITE PREPARATION METHODS

Choice of a method from those suggested should be based on your objectives, the condition of the site and the equipment available.

Manual site preparation

Manual techniques can be used to control the amount of brush logging debris (slash) and to clear suitable sites for planting over small areas.

Brush can be removed with a billhook, machete, brush cutter, or shears. Never use an ax for this type of work. The risk of accident is too great.

We suggest that you clear away brush in August; there is less chance of stump suckers developing at this time of year.

A shovel or forester's hoe can also be used to get rid of logging debris (slash) and create clear sites for planting.

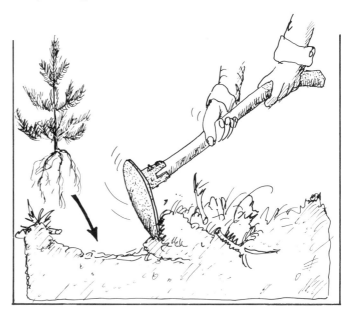

Mechanical site preparation

For larger areas mechanical techniques will make the work easier and faster.

- *Plow and harrow*

Plows and harrows are used for site preparation on abandoned agricultural land. Plowing, followed by harvesting can be carried out either over the entire site or only in places where seedlings are to be planted.

- *Brush rake*

The brush rake is a toothed blade that is mounted on the front of a crawler tractor or skidder. It clears away logging debris (slash) and piles it into windrows. It may also scarify the soil slightly.

This type of equipment is used primarily in areas where there is a considerable amount of cutting debris and little competing vegetation, and is often used in private woodlots.

- *Straight blade and V-shaped blade*

Both the straight blade and the V-shaped can be used to clear away strips of logging debris (slash) and brush. The lower edge of the blade is a cutting edge and is very effective when well sharpened.

A crawler tractor equipped with one of these blades will cut down brush and pile it to one side along with the logging debris (slash), in somewhat the same way as a plow pushes snow aside.

When using this equipment, make sure the blade remains a few centimetres off the ground. On some sites, you must take care to avoid scalping, that is, removing the humus layer, as this may improverish the soil.

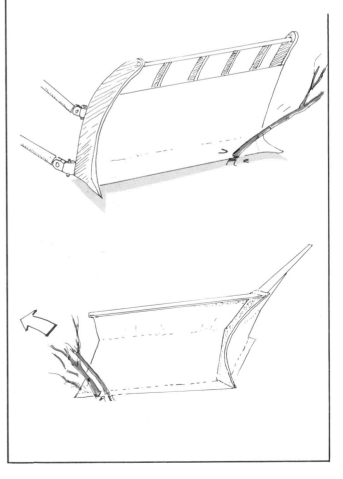

- *Disk trencher*

This is a piece of equipment used to scarify the soil. It is pulled by a skidder and is made up of two disks that till the soil as they rotate. The furrows that form as the machine advances consist of mixed soil free of logging debris (slash). Forest workers can then use these furrows as pathways.

The distance between the disks can be adjusted by either a hydraulic or a mechanical system.

- *Shark-finned barrel and anchor chain*

This piece of equipment is used primarily to scarify the soil. It consists of two toothed barrels behind which anchor chains are attached.

The positioning of the teeth on the barrels produces a rotating effect. As the machine advances, the barrels turn, forming furrows in the soil. This piece of equipment is pulled by a crawler tractor.

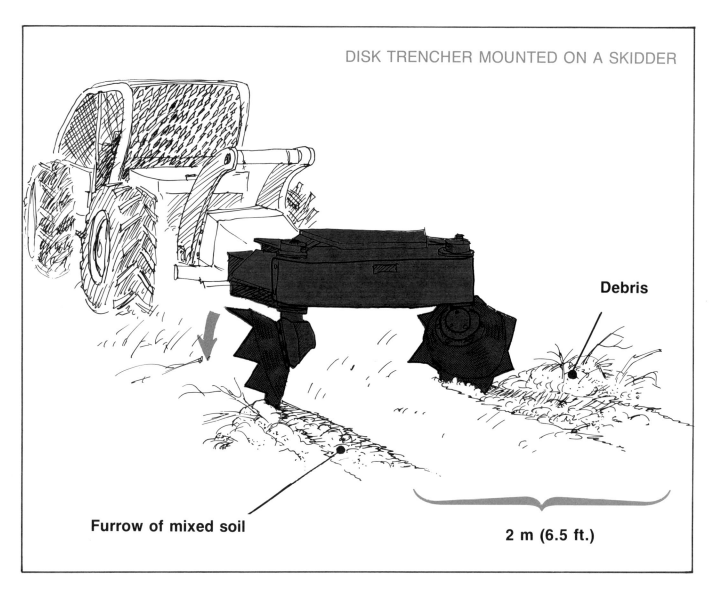

DISK TRENCHER MOUNTED ON A SKIDDER

Debris

Furrow of mixed soil

2 m (6.5 ft.)

The anchor chains, to which iron spikes are welded, rake the ground, removing logging debris (slash) and mixing the soil.

This machine must be used over a large area, because turning this machinery requires a considerable amount of space.

• *Patch scarifier*

This piece of equipment mixes the soil in small areas, leaving short furrows with gaps between them. Cutting debris is removed only where the furrows are formed.

When this tool is used, silvicultural workers will find their job more difficult. The operation does not leave cleared pathways and they must look for sites where the soil has been prepared for planting seedlings.

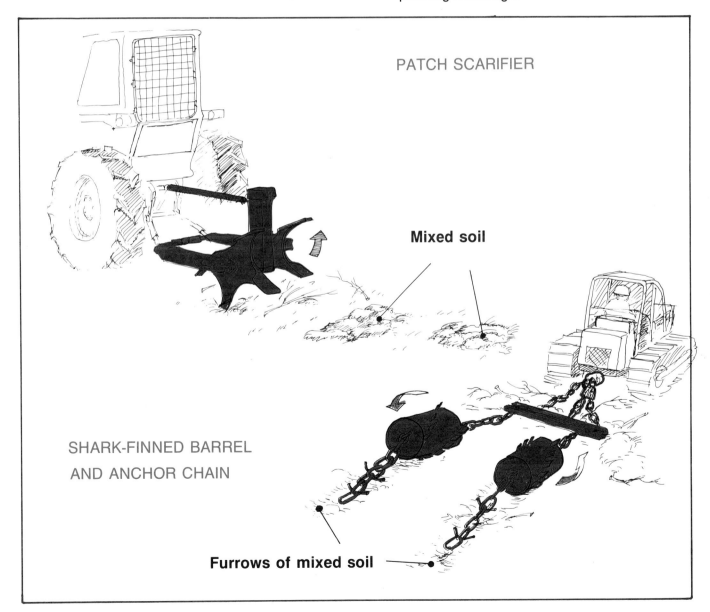

PATCH SCARIFIER

Mixed soil

SHARK-FINNED BARREL
AND ANCHOR CHAIN

Furrows of mixed soil

Chemical site preparation

Chemicals are sometimes used to eliminate grass and brush temporarily. If you are interested in this method and would like to know more about it, consult your forestry adviser, who will give you information on which products to use and how to use them.

SITE PREPARATION PATTERNS

The pattern followed in preparing the site has a considerable effect on the performance of the equipment and the progress of planting work.

The equipment should travel in straight, parallel lines wherever possible. This makes the work of the operator and the forest workers easier.

If the shape of the site is irregular, it is a good idea to work on a rectangular block first.

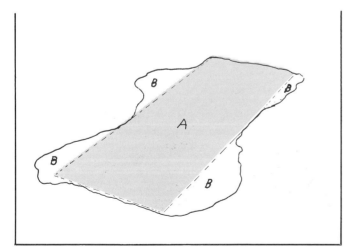

Several patterns can be used. Traveling back and forth in straight lines is effective only when the length of each line is greater than 100 m (about 300 ft.).

Remember that there must be enough space between each line to permit a seedling to be planted every 2 m (6.5 ft.).

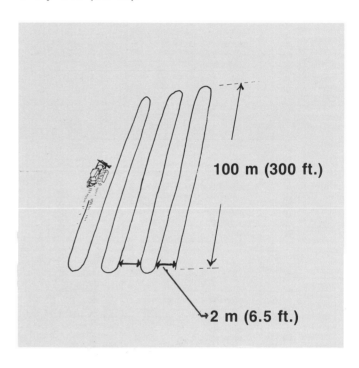

A spiral pattern considerably reduces the number of turns to be made, but makes reforestation more difficult.

Lastly, an overlapping pattern requires more organization. The site must be divided into small blocks, 20 to 40 m (66 to 130 ft.) wide. This method also takes longer for any given area, since more distance is covered.

2 m (6.5 ft.)

20 to 40 m

(66 to 130 ft.)

2 m (6.5 ft.)

WINDROWS

The positioning and length of windrows have an impact on reforestation and subsequent maintenance work. It is important to have easy access to all parts of your site, in spite of the windrows. Gaps in the windrows must therefore be created at regular intervals to permit passage.

You must therefore leave an open space through which machinery can pass, every 100 m (about 300 ft.) and at the end of each windrow, whether it is next to an access road or a wooded area.

The distance between windrows must be at least 20 m (66 ft.), and no more than 10 percent of the site's area should be taken up by windrows.

The positioning of the windrows depends on the site's topography.

- **On flat terrain,** windrows should be perpendicular to the access road.

- **On sloping terrain,** windrows should be perpendicular to the slope. The windrows will thus form a barrier, slowing down the rate of water flow and reducing the risk of soil erosion.

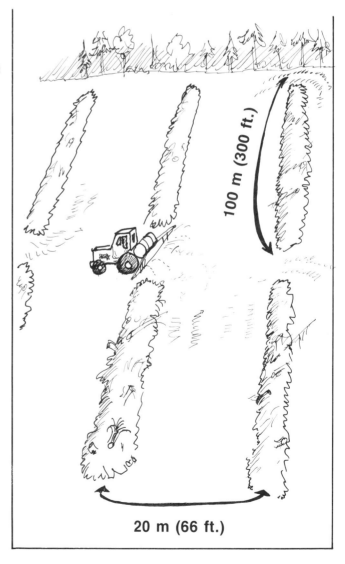

100 m (300 ft.)

20 m (66 ft.)

Summary

There are several ways to prepare a site. Some are intended to eliminate competing vegetation temporarily. Others create an environment favorable to the development of seedling roots. In any event, remember that site preparation is an important step in the management of your forest and must not be omitted because it helps ensure the survival of your seedlings.

PLANTING

PLANTING

Reforestation is a common silvicultural practice. Like other practices, it must be done carefully, using the proper tools and methods. Using the techniques described here will in large part determine how successful you are.

PLANTING

Before beginning reforestation of your site, you should be aware of the various steps essential to success. Checking, transportation, conservation, and planting are all elements with which you must be familiar in order to get the results you want.

Do yourself a favor...

First read the section on site preparation. It explains the importance of clearing away logging debris, brush, and grass. Good site preparation will temporarily reduce competition from weeds and allow the seedlings to grow better. Without such preparation, the brush will dominate the seedlings and deprive them of sunlight to the point of jeopardizing their growth.

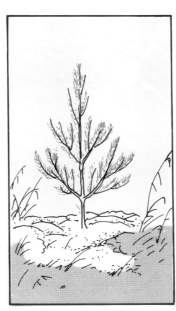

Don't forget that preparing the ground beforehand will make your planting work easier and faster.

Plan ahead

In the future, you must be able to move around freely inside your plantation. So it is important to provide access roads and firebreaks.

There should be a cleared strip 3 m (10 ft.) wide between the plantation site and adjacent roads or forest. This is an elementary precaution against fire. Similarly, there should be access roads 3 m wide at regular intervals to facilitate future maintenance and harvesting.

Decide which areas are not to be reforested: roads, areas subject to periodic flooding, rocky outcroppings, and hydro-line corridors.

PROVIDE ACCESS ROADS AND FIREBREAKS
3 M WIDE

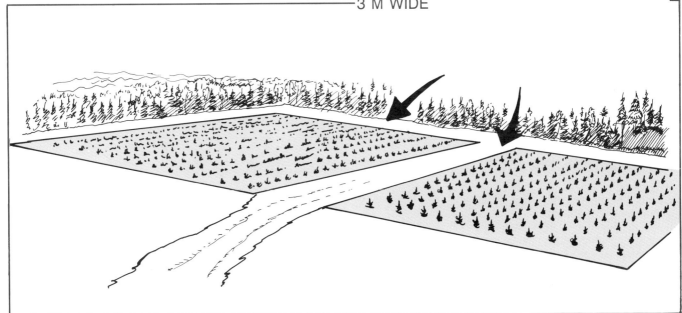

Remember that...

Bare-root seedlings are seedlings grown directly in the soil. They are usually over two years old and their roots are free when they are being replanted.

Containerized seedlings are generally grown in containers in sheltered locations (greenhouses, shadehouses). They are two years old or less and their roots are encased in a soil plug.

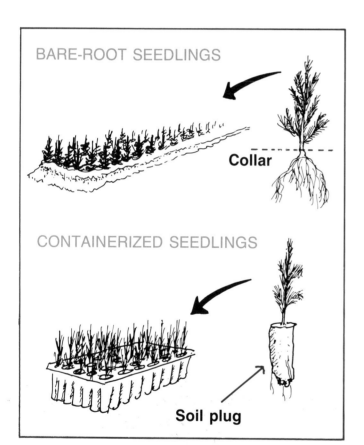

BARE-ROOT SEEDLINGS

Collar

CONTAINERIZED SEEDLINGS

Soil plug

Seedling delivery

Arrange for delivery of your seedlings during the period when reforestation has the best chance of success. As a general rule, in our climate, bare-root seedlings should be planted between May 15 and June 15. Containerized seedlings can be planted later, but in neither case should they be planted during hot, dry periods.

When your seedlings arrive, check the quality and quantity. Keep the tags that come with them.

The foliage should be healthy and well developed, the roots should be numerous and mold-free. The buds of bare-root seedlings should not be open.

When bud break has occurred, the needles of a seedling are pale green at the tip

The soil plugs of containerized seedlings should be firm and the roots well spread out. The soil plugs should not break when you remove the seedlings from the container or handle them.

Take care of your seedlings

Ideally, planting should begin the same day the seedlings are delivered. In any case, take all necessary precautions to ensure the survival of your seedlings.

Storing of bare-root seedlings to be planted within two days

- Leave the seedlings in their packaging (waxed paper, jute cloth);

- Place the bundles in the shade;

- Water the bundles daily and put them back on their sides after watering.

Storing of containerized seedlings

- Put the containers side by side on the ground in a cool place sheltered from the wind;

- Water them frequently to keep the soil plugs moist;

- You may hang a tarpaulin over the containers for shade, but never cover them so tightly as to prevent air from circulating.

Trench-storing of seedlings to be kept more than two days

- Find a shady, cool place;

- Open the bundles and place the seedlings side by side in a previously dug V-shaped trench;

- Cover the roots with moist earth;

- Water them daily.

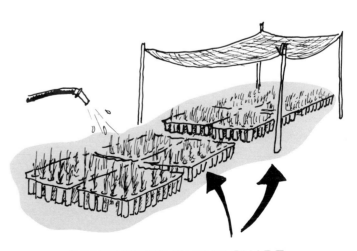

KEEP CONTAINERS IN THE SHADE

37

Getting them there

When the seedlings cannot be delivered to the plantation site, you must arrange transportation. A tractor or all-terrain vehicule, with a trailer attached, is sometimes indispensable.

Cover the seedlings with a tarpaulin during transportation to prevent them from drying out.

Once the seedlings are at the site

Bare-root seedlings:

- Carry them in a container (bucket or bag) with wet peat moss in the bottom;

- Use the peat moss from the bundles and, if necessary, add more;

- Do not expose the roots to the wind or sun.

- You can carry the container by wrapping a pair of straps around it or hanging it from two S-shaped hooks attached to your belt.

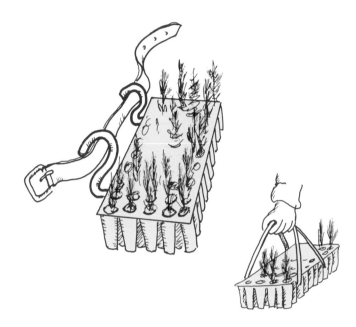

Peat moss

Containerized seedlings:

- Carry the containers in a basket or a carrier designed especially for the container;

Be prepared

- Have enough equipment (tools, buckets, carriers) for everyone;

- Distribute the seedlings (bundles or containers) around the site to minimize carrying distances;

- If the seedlings are stored in trenches, take only a few at a time to avoid drying out the roots.

Keep your distances

When reforesting, maintain the alignment and spacing of the seedlings to keep the rows straight. This will make moving through the plantation easier.

In softwood (pine and spruce) plantations, the distance between individual seedlings and rows of seedlings is 2 m (6.5 ft.). It is easier to maintain the correct distance between rows when working in pairs.

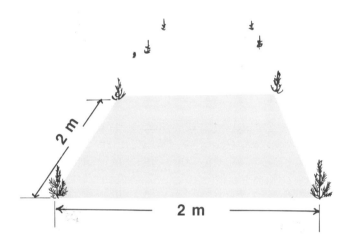

The best laid plants...

Be careful with the seedlings during planting. Whatever method and species you use, follow these basic rules:

1. Protect the roots:
   - spread them out well and never roll them up in the soil;

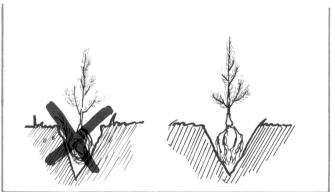

- do not break the soil plugs;

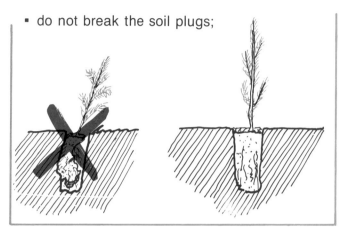

2. Never plant more than one seedling in the same hole;

3. Plant the seedlings as upright as possible (maximum incline of 10°);

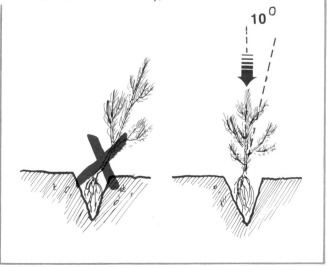

4. Plant the seedlings at the right depth:

   – the collars of bare-root seedlings should be at ground level;

   – the soil plugs should be 1 or 2 cm below ground level;

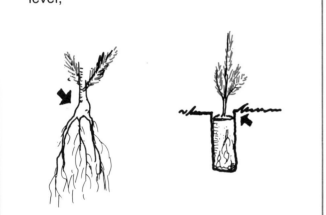

5. Never leave the roots exposed or bury the branches;

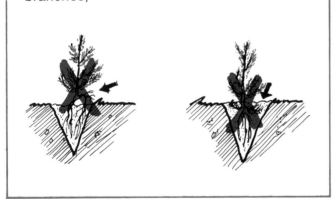

6. Pack the soil down well with your heel to avoid air pockets, which can kill the roots;

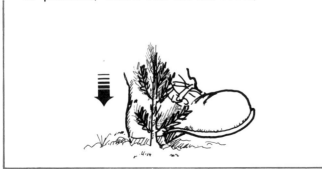

7. Do not plant seedlings in unsuitable areas (where there are water holes, stumps, or rocks). In scarified sites, plant the seedlings on the edge of the furrows;

8. Leave 2 m (6.5 ft.) between individual seedlings and between rows of plants;

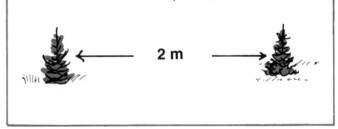

9. Never plant too closely to a natural seedling.

Protect yourself

Once the planting is finished, it is important to mark the plantation prominently with signs and to prohibit access to all-terrain vehicles and snowmobiles. Keep domestic animals off the plantation. In short, do not squander your investment and make sure you maintain your plantation.

We suggest...

Four methods of planting seedlings. Whichever method you choose, it is important to follow the basic principles.

"T" METHOD
(bare-root seedlings)

Used on agricultural land covered with thick grass or for seedlings with numerous roots.

1. Push the shovel straight down into the ground;

2. Insert the shovel straight down a second time to put the bar on the "T";

3. Press the shovel handle down toward you to open the first slit and keep the shovel in the ground at an angle;

4. Slide the plant into the opening, then pull it up until the collar is at ground level and the roots are pointed downward;

5. Withdraw the shovel and let the slit close, keeping an eye on the height of the collar; and

6. Pack the soil down with your heel, taking care not to damage the bark or the branches.

CLEFT METHOD
(bare-root seedlings)

*Used in loose (sandy, clayey) soils where
it is easy to work with a shovel.*

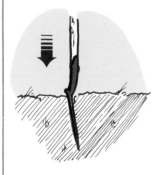

1. Push the shovel straight down into the ground;

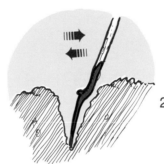

2. Pull, or push, the shovel to create an opening in the soil;

3. Place the plant in the opening, making sure that its roots are well spread out and pointed downward;

4. Adjust the height of the collar so that it is at ground level; and

5. Withdraw the shovel and pack the soil down with your heel, taking care not to damage the bark or the branches.

We recommend using a forester's shovel or a round shovel with the blade cut for these two methods.

FORESTER'S PLANTING HOE METHOD
(bare-root and containerized seedlings)

Used in various types of soil, except rocky ground.
Particularly useful on steep terrain.

1. Push the hoe blade into the ground;

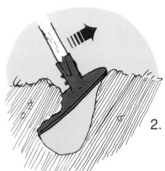

2. Pull the hoe handle upward to make an opening;

3. Holding the soil back with the hoe, place the seedling upright in the hole;

4. Replace the soil with the end of the hoe; and

The hoe can be used to clear a planting site

5. Pack the soil down with your heel.

PLANTING TOOL METHOD
(containerized seedlings)

Used on almost all kinds of ground.
A very popular tool.

1. Insert the planting tool into the ground up to the depth gauge;

Depth gauge

2. Pull out the planting tool by turning it gently;

3. Place the seedling in the hole; and

4. Pack the soil down with your heel, being careful not to damage the seedling.

44

PLANTATION MAINTENANCE

You may have completed reforesting part of your land... but your work is not over yet! The success, and especially the profitability, of your investment will depend to a great extent on the care you put into maintaining the plantation.

PLANTATION MAINTENANCE

Even if you have taken great care in preparing the site and planting seedlings, you must now work at maintaining your plantation.

If left untended, it will not produce the benefits you are counting on and its yield, in terms of volume, will be no better than that of an unmanaged forest.

The seedlings you have just planted are very vulnerable and will have to compete for light, water, and nutrients. They may also be threatened by diseases or attacked by insects or rodents. If they are not given every chance to survive, they may well perish.

You must therefore practice regular maintenance to ensure that your plantation survives and produces healthy trees.

Observing the plantation

In order to determine what kind of maintenance your plantation requires, you must visit it several times and closely monitor its progress during the growing season from May to September.

Your observations will vary according to the season. In spring, you should pay particular attention to the condition of the seedlings. In summer, competing vegetation (weeds, grass, brush) should be your main concern. In fall, you should estimate the number of surviving seedlings. At all times, you should watch for insects and diseases.

It is a good idea to note your observations. This will make your job easier and help you in planning your strategies.

IN SPRING, CHECK THE CONDITION OF THE SEEDLINGS

IN FALL, ESTIMATE THE NUMBER OF SURVIVING SEEDLINGS

Observation techniques

In general, you will have to visit the whole plantation in order to identify any problems and determine their causes.

For large plantations, it is best to study sample plots. In this way, you can assess the problems without having to examine all the trees.

Select sample plots measuring 1/100 of a hectare (containing 25 seedlings) and, for each sample plot, count the number of damaged or dead seedlings. Multiply this number by 100 to calculate the number of damaged or dead seedlings per hectare.

SAMPLE PLOT

1/100 of a hectare

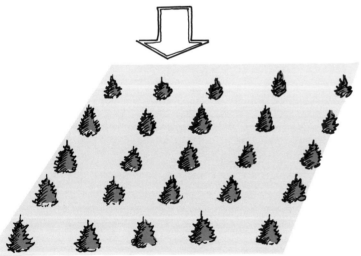

25 seedlings

For an accurate assessment, you will need to examine 5 sample plots per hectare. They should be located in areas which are most representative of the general condition of the plantation.

5 SAMPLE PLOTS PER HECTARE

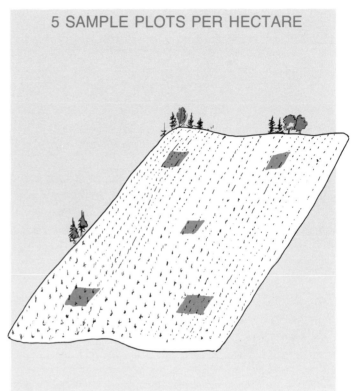

Spring maintenance

Because of our climate, spring plantation maintenance is important.

In the first years, before the crowns of seedlings have grown above snow level, the seedlings are protected from frost and winter drying by the wind. However, the weight of the snow can cause damage.

- Seedlings lying on their side: straighten them up. Where necessary, use stakes to provide support.

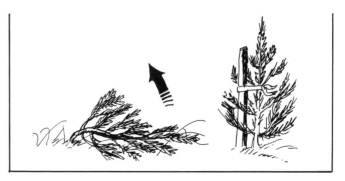

- Broken or partially detached branches: cut them off to prevent further damage.

- Multiple leader seedlings: keep the more promising stem and support it if necessary. This will prevent the development of forked trees.

CUT OFF BROKEN BRANCHES

- Broken crowns: choose the best top branch and tie it so that it grows in the direction of the stem. This is to prevent the formation of a stub.

In winter, damage can be caused to seedling roots.

Seedlings may be buried by wind in sandy soils. In such cases, the soil must be cleared to collar level.

Others may be lifted by frost or have their roots exposed by the wind (sandy soils) or water runoff. If their chances of survival are good, they should be uprooted and replanted or transplanted; if not, they should be replaced.

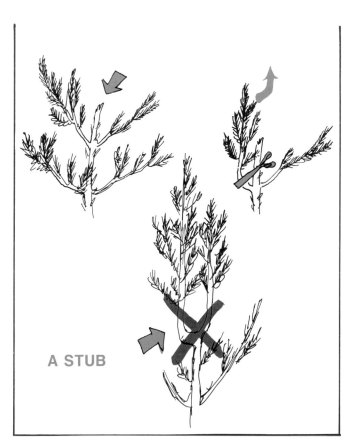

A STUB

Replacement planting

Seedlings that have died or are not expected to survive must be replaced as soon as possible. Five years after planting, there should be 2 000 healthy seedlings per hectare (800 per acre) on your plantation. This number is considered necessary for a good yield.

Replacement planting is usually done in the spring. The previous fall you will have counted the number of surviving seedlings and determined the reasons your seedlings died so that these problems can be avoided in the future.

Site conditions may be unsatisfactory (strong competition, poor drainage, compacted soil, infertile soil) in part or all of the plantation. Either the planting site must be moved over slightly or site conditions must be improved.

In containerized seedling plantations, if many seedlings have been lifted by frost, they should be replaced with bare-root seedlings.

In certain naturally regenerated areas, seedlings may not be evenly spaced. Replacement planting should be done in areas that lack seedlings.

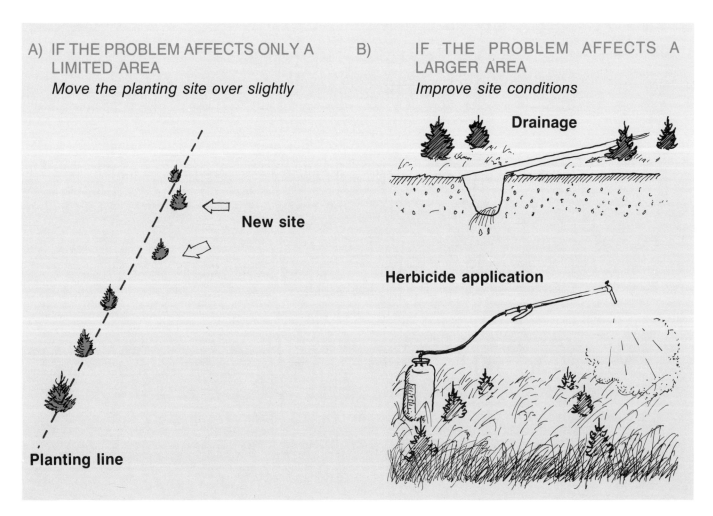

A) IF THE PROBLEM AFFECTS ONLY A LIMITED AREA
Move the planting site over slightly

New site

Planting line

B) IF THE PROBLEM AFFECTS A LARGER AREA
Improve site conditions

Drainage

Herbicide application

Controlling surrounding vegetation

It is important to control surrounding vegetation, that is, weeds and brush. Weeds (low and tall) cause problems mainly on abandoned agricultural land. Brush (mountain maple, raspberry bushes) is more often a problem in former cutting areas.

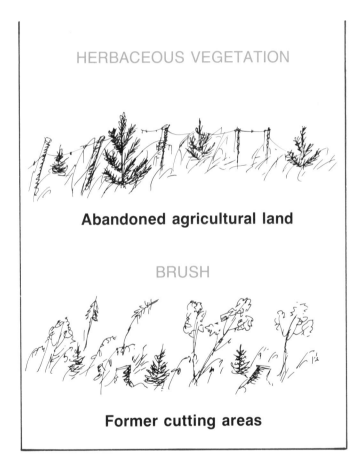

HERBACEOUS VEGETATION

Abandoned agricultural land

BRUSH

Former cutting areas

Your seedlings will survive only if this vegetation is controlled. However, this does not mean eliminating it. In fact, when surrounding vegetation is not too dense, it provides a certain amount of protection against sun scald, frost, drought, and lifting. But when it is overly abundant, it competes with seedlings for light, nutrients, and soil moisture.

Vegetation can be controlled either mechanically or chemically.

Mechanical control is undoubtedly one of the most effective ways of reducing competition. However, in the case of weeds, it has to be done several times a year. For brush, the process must be repeated until the seedlings have outgrown the brush.

Vegetation is cleared away mechanically only in the area immediately surrounding the seedlings, either in strips or around each seedling. If you choose this method, be careful not to damage the seedlings.

MECHANICAL CLEARING

In strips

Around each seedling

51

Following are some of the methods and tools used in the mechanical control of weeds and brush.

WEEDS	
Methods	**Tools**
Weeding	• forester's hoe • spade
Mowing	• scythe • billhook
Covering the ground	• mulch

BRUSH	
Methods	**Tools**
Cutting or girdling	• machete • swedish brush ax • billhook • brush cutter
Girdling	• ax or machete
Devitalization	• chemical product (brushkiller)

The ground can be covered by placing straw, wood chips, or plastic around the base of seedlings.

Girdling makes it possible to kill unwanted trees without having to cut them down.

Devitalization is the process by which a chemical is applied on stumps to avoid suckers or under bark to kill a tree.

Brush should be cleared away in August when there is less risk of stump suckers.

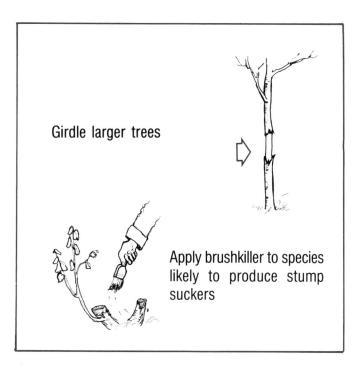

Girdle larger trees

Apply brushkiller to species likely to produce stump suckers

Chemical control

Chemicals are an effective means of controlling weeds and brush.

They can be applied over entire areas or on individual rows or seedlings. Always use the proper quantity for the size of the surface you are treating.

AREA ACTUALLY TREATED

Different types of equipment can be used for spraying. Some sprayers are portable (motorized or manual), while others are mounted on a tractor or skidder.

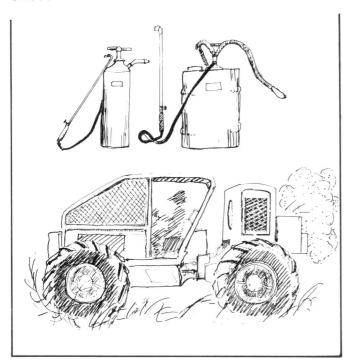

Remember that chemicals can be dangerous to your health and to the environment. Many precautions must be taken in handling them.

If you are thinking of chemically treating your seedlings, contact your forestry adviser, who will tell you which chemicals to use and how to use them.

Growth improvement

In addition to ensuring that your plantation survives, you must take steps to promote its growth and improve its quality. Various methods such as fertilization and pruning can help you achieve these goals.

Fertilization

If your seedlings do not seem healthy, or have short or discolored needles, they may not be getting all the nutrients they need from the soil.

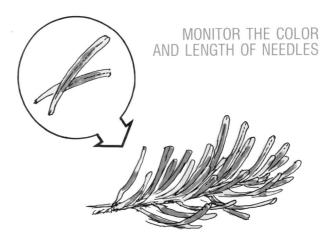

MONITOR THE COLOR AND LENGTH OF NEEDLES

If this is the case, they should be fertilized. Before doing so, however, take samples of the soil and foliage. Have them examined by a specialist, who can tell you what your plants are lacking.

Fertilization is not recommended before the plantation's third year. Surrounding vegetation benefits as much from fertilizer as the seedlings do.

Chemical fertilizers must be applied in the spring, while organic fertilizers (manure, compost) can be used at any time.

Application can be done manually or with a motorized sprayer. The fertilizer should be applied on the ground, directly under the foliage, and not on the leaves or needles. Be sure to use the recommended quantity. If a chemical is improperly applied or too much of it is used, the seedlings can be burned.

APPLY FERTILIZER DIRECTLY UNDER THE FOLIAGE AND DO NOT EXCEED THE PRESCRIBED APPLICATION RATE

Pruning

Seedlings are pruned for a variety of reasons.

- To remove low branches so as to prevent damage caused by snow or freezing rain;

- To remove branches broken by snow, freezing rain, animals, and other factors;

- To prevent the spread of disease;

- To produce wood with fewer knots, which is better for the production of lumber and veneer.

Pruning to prevent damage from freezing rain or snow should be started when the trees are about 2.5 m (8 ft.) high. If your goal is to produce wood without knots, you should wait until trees reach diameter at breast height (dbh) of 8 to 10 cm (3 to 4 in.) before pruning.

TO PREVENT DAMAGE

2.5 m
8 ft.

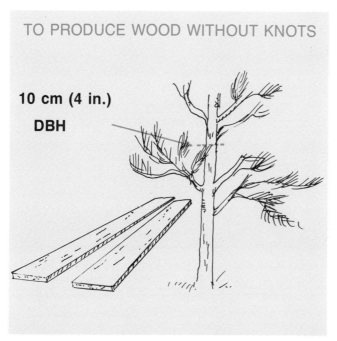

TO PRODUCE WOOD WITHOUT KNOTS

10 cm (4 in.)
DBH

Pruning should be carried out while the trees are inactive (between the end of August and the beginning of May).

No more than 1/3 of the height of a tree should be pruned. Further pruning may retard its growth.

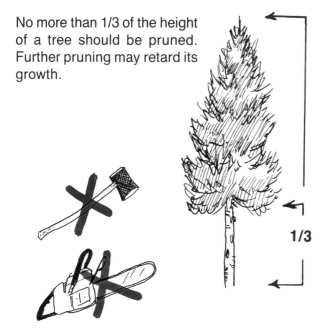

1/3

Pruning saws and pruning shears are the tools that should be used. Never use a chain saw or ax, which may damage the trees and create a point of entry for disease.

Branches must be cut off as close as possible to the trunk without cutting the swelling at the base of the branches. Branches more than 4 cm (1.6 in.) in diameter should not be removed since they will leave excessively large scars.

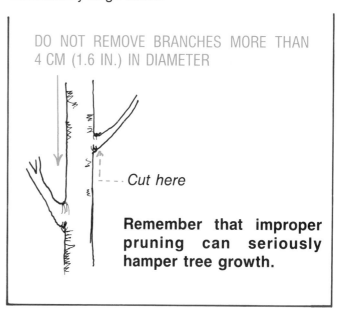

DO NOT REMOVE BRANCHES MORE THAN 4 CM (1.6 IN.) IN DIAMETER

--- *Cut here*

Remember that improper pruning can seriously hamper tree growth.

Plantation protection

Trees on plantations are threatened by more than just competition from surrounding vegetation, poor soil quality, and sometimes extreme temperatures. Other factors can cause damage as well.

Among the most harmful are insects and diseases. See the section on forest protection for guidance on how to handle these problems.

Animals, wind, fire, and people can also cause considerable damage.

Animals

Rodents (hares, field mice, porcupines, and others) can seriously harm trees by gnawing the bark or eating the foliage.

Field mice are found primarily in grassy areas. Proper control of this vegetation before the winter will limit damage.

Finally, domestic animals should be kept off plantations to prevent grazing and trampling of seedlings, and compacting of soil.

Wind

Some plantations are exposed to the wind. In winter, seedlings that are not protected by a cover of snow may dry out (winter drying).

The solution is to install snow fences (windbreaks) to reduce wind velocity and allow the snow to accumulate.

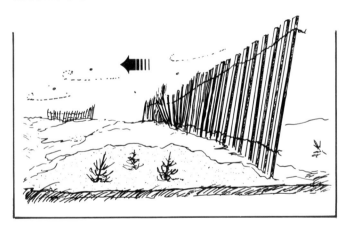

Fire

For protection against fire, 3-m (10-ft.) strips should be cleared of vegetation to act as firebreaks. In addition, a water reservoir should be created by damming a small body of water in the area.

Public access

All-terrain vehicles, snowmobiles, and people should be prohibited from entering your plantation. Appropriate signs must be put up to identify the area as a plantation.

Careful planting is essential to a succesful plantation, but it will be of little use if proper maintenance is not carried out.

Monitoring the seedlings to detect any growth deficiency or site condition problems is critical to plantation maintenance. It is also an inexpensive way to ensure that you will be able to take prompt action if a problem should arise.

If you notice certain problems in your woodlot, for example, if the soil is very wet and the trees do not seem to be growing well, forest drainage might be the answer. This section answers your questions on this silvicultural practice.

FOREST DRAINAGE

Water plays a crucial role in tree growth. It carries the nutrients essential to development.

This water is absorbed by the roots. To achieve maximum absorption, the roots also need oxygen.

If the soil is saturated with water, the roots cannot breathe. They develop poorly and remain near the surface of the soil. As a result, the trees do not grow well and have little resistance to windfall.

What is a poorly drained site?

A site is poorly drained if the water level is close to the surface of the soil during a good part of the growing season, namely May to September.

Causes of poor drainage

Soil drainage is often more deficient on flat land.

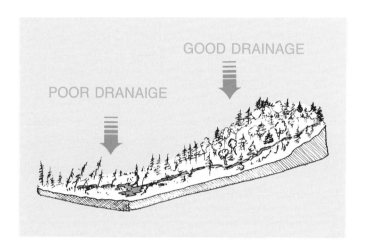

Inadequate drainage can also be caused by an impermeable layer in the soil, which prevents water from penetrating.

Forest drainage

Forest drainage is a management technique designed to improve growth conditions for woodlot trees.

In farming, land drainage is a common practice. Though less frequent, forest drainage is done for the same reasons as agricultural drainage. It makes it possible to:

- drain off surface water;

- lower the water level in the soil;

- improve soil aeration;

- make room for healthy root development; and

- improve soil fertility by promoting better decomposition of organic matter.

Soil is made up of humus, sand, gravel, and many small cavities called pores. When drainage is adequate, the pores are filled with air and water. If the soil is poorly drained, the air in the pores is replaced by water. The purpose of drainage is to restore the balance between water and air in the soil.

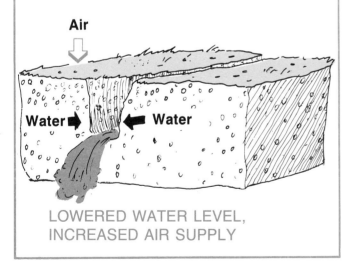

Air

Water ➡ ⬅ **Water**

LOWERED WATER LEVEL, INCREASED AIR SUPPLY

After drainage has been done on the wet areas of your woodlot:

- Your woodlot yield will increase;

- Your trees will have better annual growth;

- You will be able to use sites that were considered non-productive; and

- Your woodlot will thus increase in value.

Drainage should not, however, cause the soil to dry out. Too little water in the soil is just as harmful to tree growth as too much water.

Which stands to drain

The choice of sites to drain should be based on which stands are most likely to respond well.

In eastern Canada, the stands that are most likely to respond well to drainage are those containing:

- *black spruce;*

- *red spruce;*

- *balsam fir;*

- *white spruce;*

- *cedar;*

- *red maple; and*

- *poplar.*

For other regions in Canada, please consult your registered professional forester.

Drainage can be done on such lower-yield sites as peat bogs, but you should concentrate primarily on forest sites that will respond immediately to treatment.

Ditch drainage is one method of increasing the yield of stands in wet soil, where growth is limited.

Ditch network

Drainage is done by means of a network of ditches extending over the drainage site.

Before work is started, the site must be studied to determine an economical and effective drainage system. Consult a forestry adviser qualified to conduct such a study.

The first step is to assess the slope and size of the site and the type of soil.

Before digging, clear an area of land 5 to 6 m (16 to 20 ft.) wide where the ditches will be dug. Machinery will thus be able to circulate freely, dig the ditches, and move material without damaging trees.

5 to 6 m (16 to 20 ft.)

The excavated soil should be deposited in discontinuous piles on either side of the ditch so that water from precipitation can drain off by flowing into the ditch.

Proper planning of the ditches will prevent erosion and premature deterioration.

In general, vegetation will begin to grow on the ditches' embankments two to three years after the ditches have been dug. However, in soil with rather coarse sand, it is preferable to speed up the growth of embankment vegetation. More rapid stabilization of the soil will reduce erosion.

Main ditch

The main ditch directs water to an existing ditch or a natural watercourse. It always runs in the same direction as the slope and is located in the lowest part of the drainage site.

The size of the main ditch varies from one location to another, depending on the volume of water to be drained off.

A sedimentation basin should be dug at the end of the main ditch.

The sedimentation basin will slow down the flow of water and collect suspended particles, which are deposited on the basin floor. Therefore, water quality of natural watercourses is not affected by the residue in water from the ditch network.

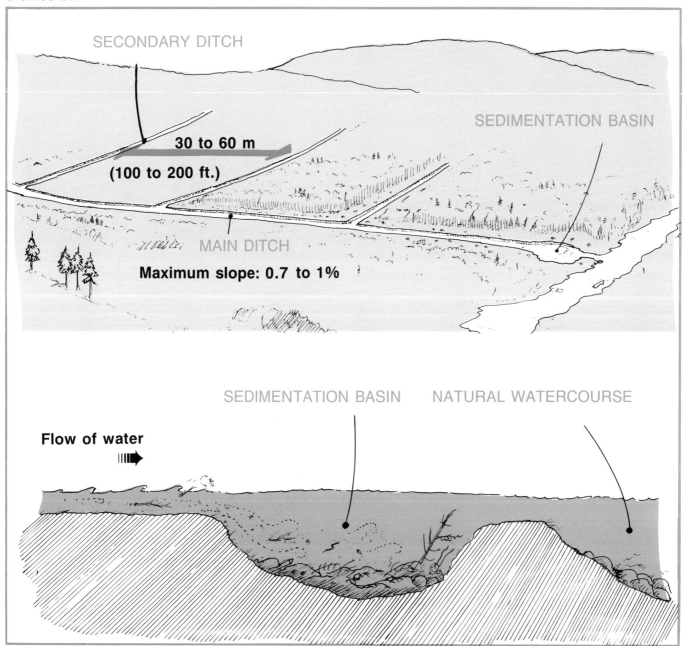

SECONDARY DITCH

30 to 60 m

(100 to 200 ft.)

SEDIMENTATION BASIN

MAIN DITCH

Maximum slope: 0.7 to 1%

SEDIMENTATION BASIN   NATURAL WATERCOURSE

Flow of water

For the sedimentation basin to be effective, it must be cleaned regularly by removing the sediment (particle) deposits from it.

The main ditch should have a slope of 0.7 to 1%. If the slope is too steep, the ditch walls will erode and a large quantity of sediment will be carried in the water.

Secondary ditches

Secondary ditches run perpendicular to the general slope of the site and direct water into the main ditch.

Their purpose is to collect run-off water from the surface of the soil and water contained in the soil.

When the soil is rich in humus (black soil), the secondary ditches should be U-shaped.

If the soil has a higher sand content, the ditches should be V-shaped. The gentler embankment slope will minimize cave-ins.

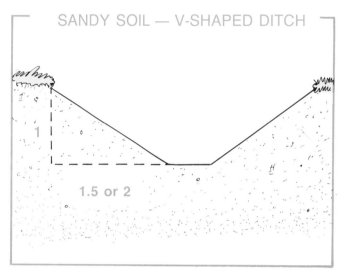

SANDY SOIL — V-SHAPED DITCH

The distance between the ditches depends on the volume of water to be drained, the type of soil and the size of the stand.

You should plan to dig secondary ditches 30 to 60 m (100 to 200 ft.) apart.

They may be 60 to 90 m (2 to 3 ft.) deep and 90 to 140 cm (3 to 5 ft.) wide at the top.

HUMUS-RICH SOIL: U-SHAPED DITCH

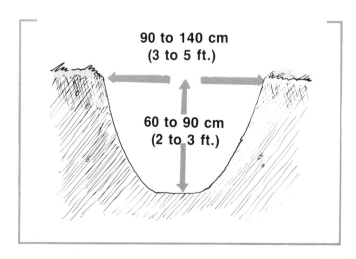

90 to 140 cm
(3 to 5 ft.)

60 to 90 cm
(2 to 3 ft.)

Network contour ditch

It is sometimes necessary to have a contour ditch, which is designed to collect water from outside the drainage site.

The contour ditch should be the same size as the secondary ditches.

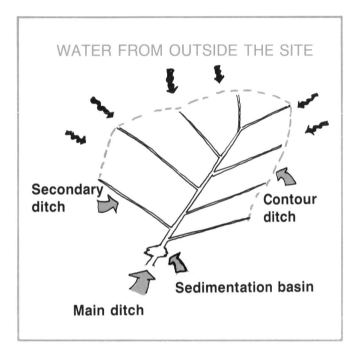

Ditch maintenance

To keep the ditches working properly, they must be cleaned periodically. A yearly inspection of the ditches will reveal landslip (mostly sand) and a build-up of vegetation that should be removed. Twenty years after the ditches have been dug, the bottoms will have to be redug. If the soil is sandy, this operation may have to be done sooner.

The sedimentation basin also needs to be maintained and cleaned. The cleaning must be done more often.

Equipment used

Forest drainage is done mainly with a backhoe mounted on a crawler tractor or tires.

A farm tractor fitted with a backhoe may also be used.

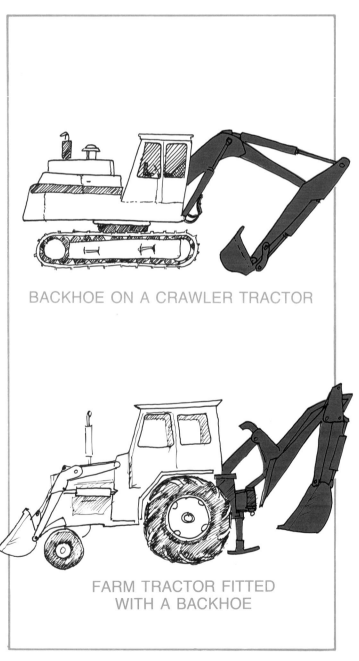

Whatever equipment is used, the backhoe will usually have a U-shaped or V-shaped bucket, depending on the type of soil on the drainage site. A bucket of the right shape and size speeds up the work.

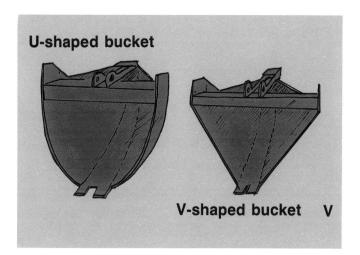

U-shaped bucket

V-shaped bucket   V

- the description and size of the drainage site.

To be valid, this property-owner's agreement must be notarized, or approved by the property owners of at least 3/4 of the planned drainage site.

Before undertaking the work, the property owner(s) must submit the plan for approval to the municipal office of every municipality in which all or part of the main watercourse will be located.

Usually, it is necessary to apply to your provincial department handling environmental affairs for an authorization certificate. The application includes the description and location of the work and enables authorities to ascertain whether the work would affect a drinking water source.

Agreements

Watercourses created by forest drainage are mainly municipal watercourses that are neither navigable nor floatable, and governed by the municipal code.

If several property owners are involved in the drainage work, they must reach an agreement, which may take the form of a by-law, surveyor's report, or deed of agreement. The deed of agreement is the most commonly used method.

A deed of agreement must include:

- the description and location of the watercourse;

- specifications for the work to be done;

- the ways in which the property owners will contribute to the work; and

If the drainage site is larger than 25 km² (10 mi.²), an impact study is required.

The main steps in carrying out forest drainage are:

1. Observation of a problem by one or more property owners;

2. Agreement (deed of agreement) among the property owners involved;

3. Consultation with a forestry adviser;

4. Site study for the purpose of preparing plans and specifications;

5. If required by your province, application to your provincial department handling environmental affairs;

6. Filing of drainage plan and deed of agreement with the municipal council and any other agency concerned;

7. Invitation to tender and contract bids; and

8. Delineation of ditch locations. Clearing and digging.

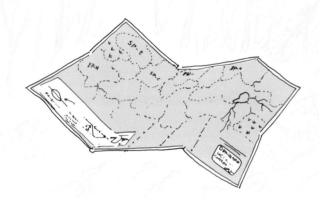

CUTTING

The primary objective of cutting is to harvest a stand with a view to its regeneration. Various methods can be used. It is essential to choose the one which best suits the conditions of your woodlot. Each of the different methods is described on the following pages.

CUTTING

Forest management includes the cutting of trees in order to:

- *harvest mature trees;*

- *encourage stand renewal;*

- *provide the best conditions for tree growth;*

- *eliminate unwanted trees (such as diseased trees).*

Various methods are used, each suited to particular conditions.

Some cutting methods are intended to improve stand quality. These methods are known as thinning and are discussed separately in this guide. Other methods are used to harvest mature trees while ensuring regeneration of the site. This is known as regeneration cutting. Still other operations, such as conversion cutting and sanitation cutting, are intended to rectify specific or unusual situations.

REGENERATION CUTTING

As the name indicates, regeneration cutting promotes development of a new stand.

Each of these methods is designed for specific conditions. A poor choice can threaten the chances of regeneration.

NATURAL REGENERATION ALREADY UNDER WAY IN A CONIFER STAND

Uneven-aged stand

This is a stand containing trees of different ages forming at least three stories: an overstory, an intermediate story and an understory.

Selection cutting is used to maintain the uneven-aged character of this type of stand.

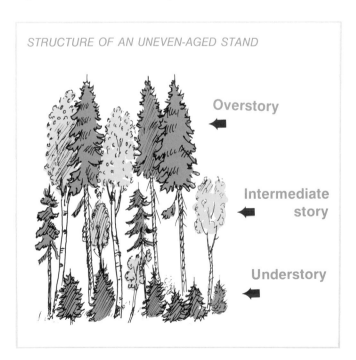

STRUCTURE OF AN UNEVEN-AGED STAND

Overstory

Intermediate story

Understory

The selection method involves repeated cutting activities in a stand. On each occasion, a small quantity of mature trees is harvested. These mature trees are replaced by saplings, which already existed at the time of cutting, or which spring up in the openings created.

Overly dense areas of saplings are thinned out at the same time as the mature trees are harvested. Selection cutting thus combines regeneration cutting with improvement cutting, intended to provide the remaining trees with better growing conditions.

In selection cutting, poor-quality trees (forked, damaged), diseased trees (canker, rust), and mature trees are harvested. No more than 10% of the stand's volume should be removed at any one time.

Selection cutting encourages the establishment of shade-tolerant species such as maple and fir.

Even-aged stand

An even-aged stand consists of trees which are all about the same age and form a single story.

The best example of this type of stand is a plantation. All the trees are about the same age and can be treated in the same way at the same time.

Regeneration methods used in even-aged stands include:

- clear cutting;

- clear cutting over small areas;

- shelterwood cutting.

These techniques are used to preserve the even-aged character of the stand.

Clear cutting

Clear cutting has the advantage of allowing a mature stand to be completely harvested in one operation. All trees of merchantable size (diameter at breast height of more than 10 cm (4 in.)) are removed.

This cutting method is recommended only if regeneration of desirable species has already begun.

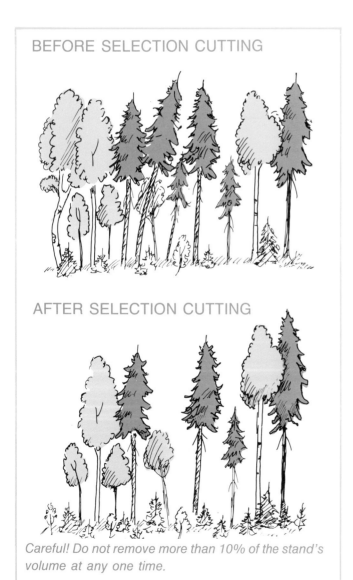

BEFORE SELECTION CUTTING

AFTER SELECTION CUTTING

Careful! Do not remove more than 10% of the stand's volume at any one time.

Regeneration is considered to be well under way if at least one seedling of a desirable species is found every 2 m (6.5 ft.).

Desirable species are those used in industry, such as fir, spruce, maple, and yellow birch.

- It increases the risk of soil erosion and fire and can lower the quality of life for the region's wildlife;

- It must be followed by reforestation if natural regeneration is not under way or is insufficient.

AT LEAST ONE SEEDLING EVERY 2 m (6.5 ft.)

CLEAR CUTTING ENCOURAGES SOIL EROSION AND REGENERATION OF UNWANTED SPECIES

Clear cutting does, however, have some major disadvantages:

- It encourages the development of unwanted species (such as raspberry and mountain maple);

In stands where regeneration is not well under way, one of the following two methods is preferable.

Clear cutting over small areas

Clear cutting over small areas works well in private forests and promotes rapid development of natural regeneration. This cutting method is therefore used in places where regeneration has not yet begun.

Clear cutting over small areas can take different forms, but each method ensures that:

- trees which will produce the seeds needed for regeneration are left near the cutting site;

- the wildlife habitat is preserved;

- the risk of soil erosion is reduced.

Different methods of clear cutting over small areas

• Clear cutting in strips

This method consists of dividing the area to be regenerated into groups of three strips. The first strip is cut, then about five years later, when regeneration is well under way, the second strip is harvested. Five years after that, the last strip is harvested.

The strips have the following characteristics:

- **Width:** 20 m (66 ft.);

- **Position:** on flat terrain, strips should be perpendicular to the prevailing winds. However, if the remaining trees are likely to be toppled by the wind (windfall), the strips must be positioned in the direction of the prevailing winds. On sloping terrain, the strips should be in the direction of the slope.

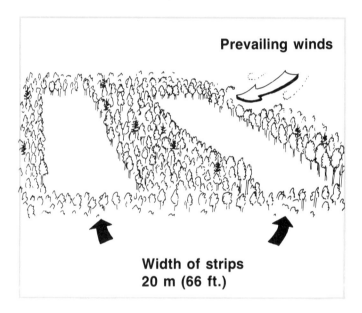

**Width of strips
20 m (66 ft.)**

• Clear cutting in a checkerboard pattern:

This method of harvesting timber gives the stand a checkerboard pattern. During the first harvest, trees are felled in a number of blocks. When regeneration is well under way, in about five years, the remaining blocks of trees are harvested.

• Clear cutting in patches:

This cutting method consists of harvesting timber over a smaller area, without following any precise pattern. Clear cutting in patches is frequently used in private forests where, because stands are often small, it can be easily done.

Shelterwood cutting

Shelterwood cutting consists of carrying out a series of partial cuttings which gradually open up the stand, so as to promote regeneration.

Successive cuttings encourage the production of seeds by the residual trees and allow more light to reach the ground.

There are three main stages in shelterwood cutting.

1° Preparatory cutting

Preparatory cutting promotes development of the crowns of the residual trees and encourages the production of seeds.

This first cut is used to remove unwanted species, diseased or malformed trees, and trees with small crowns. The volume of wood removed should not exceed 15% of the standing volume. The healthiest trees, which will produce the best seeds, are left standing.

Preparatory cutting is not necessary if the stand has already been thinned.

2° Seeding cutting

Seeding cutting, done a few years later, promotes the establishment and development of seedlings.

Between 30 and 35% of the standing volume is removed to open up the stand.

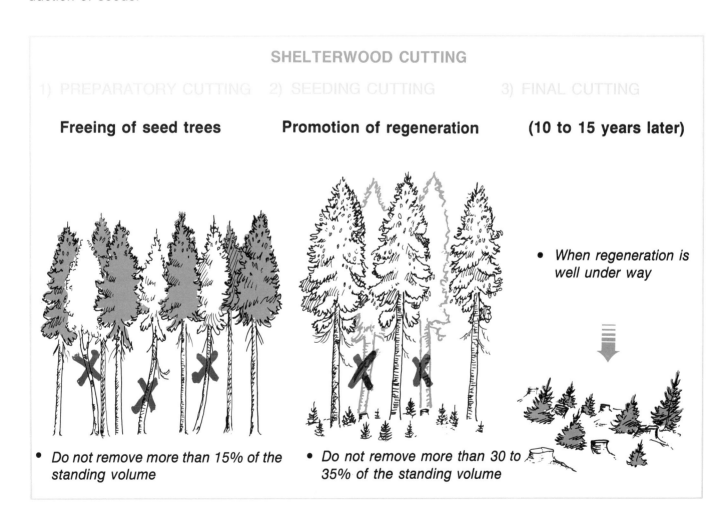

SHELTERWOOD CUTTING

1) PREPARATORY CUTTING   2) SEEDING CUTTING   3) FINAL CUTTING

Freeing of seed trees   **Promotion of regeneration**   **(10 to 15 years later)**

- *When regeneration is well under way*

- *Do not remove more than 15% of the standing volume*

- *Do not remove more than 30 to 35% of the standing volume*

3° Final cutting

The final cutting is done when regeneration is well under way, usually five to ten years after the first cut. All merchantable trees are then harvested.

The three cuttings must be carried out at fairly close intervals so that the stand will retain its even-aged character.

The primary advantage of this cutting method is the promotion of abundant regeneration in desirable species. However, one disadvantage to this method is that some of the regeneration is destroyed when the final cutting takes place.

OTHER CUTTING METHODS

Other cutting methods are designed to rectify specific situations. They can help restore productivity to a stand, improve its quality, or create a new stand.

Three cutting methods are covered under this heading.

Conversion cutting

Conversion cutting is the first step in restoring productivity to a deteriorated and unpromising stand. This method is used in stands that have been subjected to improper cutting at overly frequent intervals or have experienced repeated onslaughts of insects or diseases. Often, the volume of such stands is small, the trees are of poor quality and are growing slowly, and there is an absence of natural regeneration.

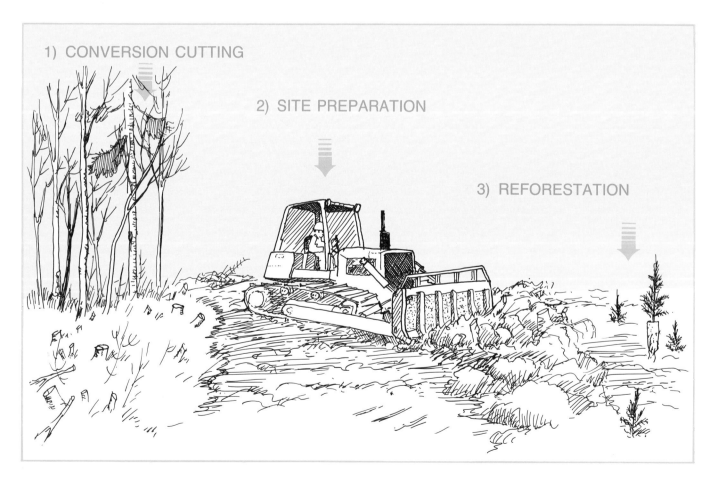

1) CONVERSION CUTTING

2) SITE PREPARATION

3) REFORESTATION

In such cases, all the trees are harvested, including those with commercial value. This procedure facilitates site preparation and reforestation at the same time.

Succession cutting

The purpose of succession cutting is to harvest a mature stand hampering the growth of a young stand at a lower story.

This cutting operation can be used to promote the development, in terms of both quality and quantity, of a young stand comprising desirable species.

For example, this procedure is often used in mature poplar stands where regeneration in fir and spruce is taking place.

When performing succession cutting, special care must be taken to select a harvesting technique which will not damage the understory, since the goal is to promote its development.

Sanitation cutting

Sanitation cutting is used to ensure that a stand is cleared of trees killed or weakened by diseases or insects, in order to lessen the risk of these spreading.

This is a preventive cutting method which must be used only if the damage is limited and the risk of spreading is significant.

SUITABLE STAND FOR SUCCESSION CUTTING

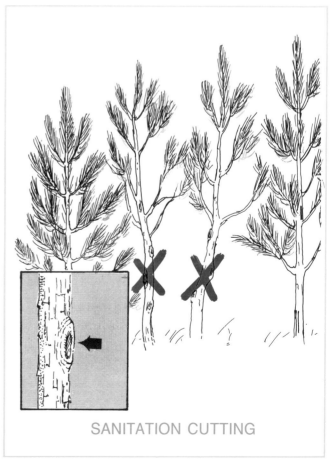

SANITATION CUTTING

Conclusion

Not all of these cutting methods are appropriate for all sites. You must know your stand well in order to determine the cutting method best suited to its condition and your objectives.

Lastly, plan road placement carefully, since some cutting methods require a number of return visits to the site.

EVEN-AGED STAND

REGENERATION   NO REGENERATION

- **Clear cutting**

- **Clear cutting over small areas**
- **Progressive cutting**

UNEVEN-AGED STAND

- **Selection cutting**

Thinning is a very common silvicultural practice that improves both the growth rate and quality of trees. To obtain good results, however, you must follow certain rules.

THINNING

Thinning is a partial cutting made in overly dense stands. It is carried out in immature stands both during and after the sapling stage. The aim is to improve the stands by releasing the best trees so as to encourage their growth.

During normal development, stands change. Many trees do not reach full growth because they cannot withstand unfavorable elements such as windfall, insects, diseases, or competition for air, water, and minerals.

This woody material, which would otherwise return to the soil through decomposition, is removed during thinning. In thinning, we make use of and accelerate the process of natural selection by removing the less promising trees. Thus, we choose the trees that will remain for the final harvest.

There are two main types of thinning.

When the cutting takes place in a stand where the trees have a diameter of less than 10 cm (4 in.), the trees felled can serve no commercial purpose. This is *precommercial thinning*.

Commercial thinning is when the trees are past sapling stage and have some commercial value. Traditionally, trees with a diameter greater than 10 cm (4 in.) are considered to have market value.

PRECOMMERCIAL THINNING

COMMERCIAL THINNING

PRECOMMERCIAL THINNING

What is it?

This is an intermediate cutting done in overly dense stands at the sapling stage. These stands are generally less than 25 years old. The diameter at breast height of the trunks is less than 10 cm (4 in.). These trees should be at least 2.5 to 3 m (8 to 10 ft.) tall.

A stand is too dense when the branches of neighboring trees intertwine. When this happens, the trees are competing for light and air, two elements essential to the growth of a tree. If these are not available in sufficient quantities, the trees' growth slows down. The trees also compete for water and minerals in the soil.

Precommercial thinning reduces competition by releasing certain trees. Rather than wait for some trees to prevail, it is better to select the ones you want and provide them with enough light, nutrients, and space.

Once you have thinned the stand, the remaining trees will:

* *develop better roots; and*

* *grow faster, particularly in diameter.*

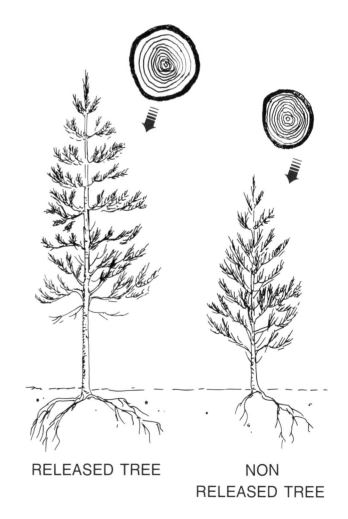

RELEASED TREE   NON RELEASED TREE

In this way, you encourage the growth of trees you select.

Your woodlot will contain better-quality and more valuable trees.

The first step is to choose the trees to release. These trees are called crop trees.

Choosing the crop trees

It is important to make the right choice of crop trees, since these will make up your final harvest.

A crop tree:

- *belongs to a desirable species (spruce, sugar maple, yellow birch);*

- *is in good health, with a straight trunk, and attractive appearance; and*

- *has grown well over the two or three preceding years.*

It is not always possible to mark the trees before thinning begins. When this is the case, you must choose while you are thinning. Take your time — mistakes cannot be corrected.

Precommercial thinning must be done according to certain rules because the aim is not to create a plantation with natural stands.

How to thin

The procedure varies depending on the species in the stand. However, the steps are the same for all types of stands.

First, decide which species you want to release. Experience has shown that it is risky to preserve only one species. You may favor one species, but always keep some trees of other species. Diversity of species in a forest makes it less vulnerable to such natural enemies as insects or diseases.

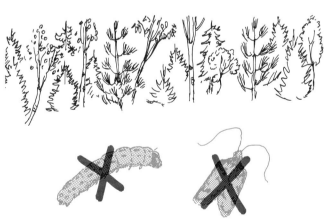

For example, you may decide to favor yellow birch, but that does not stop you from leaving some maples, white birch, spruce, or even a few fir trees.

Decide on the spacing to leave between crop trees. The spacing varies depending on the species chosen.

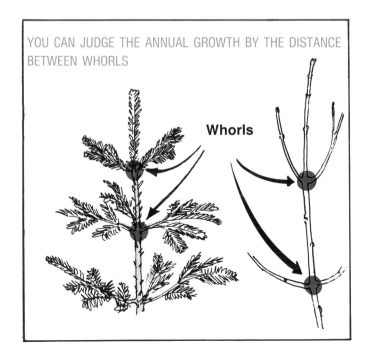

YOU CAN JUDGE THE ANNUAL GROWTH BY THE DISTANCE BETWEEN WHORLS

Whorls

The extent to which you release crop trees depends on their tolerance of shade. Trees that like shade (spruce, sugar maple) can be left more cover than those that prefer light (yellow birch, pine).

TOLERANT SPECIES

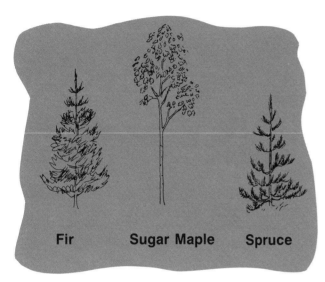

Fir **Sugar Maple** **Spruce**

INTOLERANT SPECIES

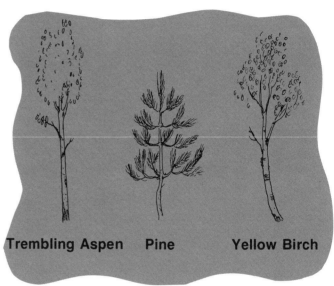

Trembling Aspen **Pine** **Yellow Birch**

DON'T CREATE A PLANTATION

ALWAYS KEEP SOME COMPANION SPECIES

Softwood stands

Precommercial thinning should be done in stands less than 25 years old when the branches of neighboring trees are heavily intertwined.

Crop trees should be spaced about 2 m (6.5 ft.) apart.

Remove all the trees immediately surrounding the crop tree and any other trees that might be harmful to it.

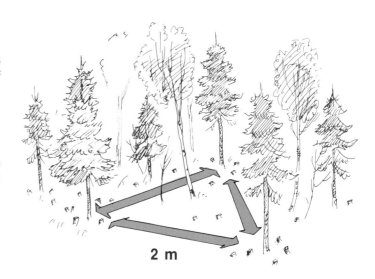

2 m

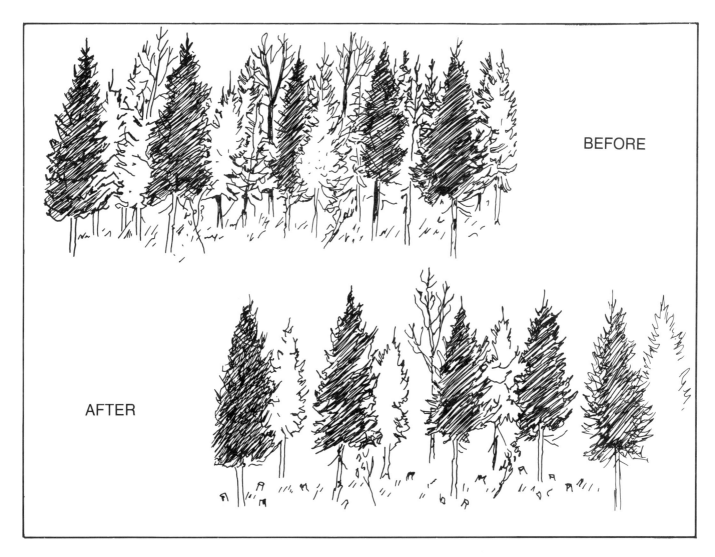

BEFORE

AFTER

Hardwood stands

Hardwood stands should still be dense after precommercial thinning. Crop trees must be allowed to increase in height without developing lateral branches.

Thin stands between 10 and 25 years old in which the branches of the trees are intertwined.

If you can choose crop trees, pick one every 5.5 m (18 ft.). Remove all the trees that overshadow the crown of the crop tree within a radius of 60 cm (2 ft.).

Release the crowns of crop trees. Do not remove the other trees! They prevent the crop trees from developing lateral branches by reducing the amount of light that reaches their trunks.

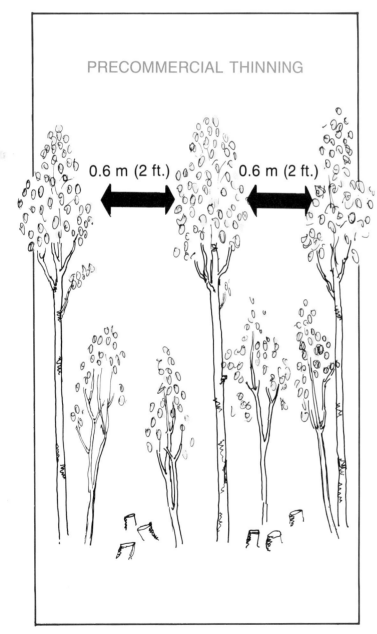

PRECOMMERCIAL THINNING

0.6 m (2 ft.)   0.6 m (2 ft.)

Crop tree (1 every 5.5 m)

- ELIMINATE TREES THAT OVERSHADOW THE CROWN

- KEEP TREES THAT OVERSHADOW THE TRUNK

If you find it difficult to select crop trees, proceed systematically by releasing a tree every 2 m (6.5 ft.).

SELECT A CROP TREE EVERY TWO METRES

Be careful

Before beginning to thin, check for insects in the stand. If the stand is infested, you must be very cautious when thinning. If there is an infestation, consult your forestry adviser.

Eliminate all trees with signs of diseases (canker, rust) as well. This prevents diseases from spreading.

Tools

There are various tools you can use for precommercial thinning. The tools fall into two groups: hand and power tools.

The hand tools are the machete, swedish brush ax, and pruning shears. These tools are used for small areas. They are not very efficient and require a lot of effort. They can be used to cut or girdle trees.

Machete

Swedish brush ax

Pruning shears

GIRDLING LETS YOU KILL BIGGER TREES WITHOUT HAVING TO CUT THEM DOWN

The power tools are the chain saw and the brush cutter.

You can work more quickly with a chain saw than with hand tools. Therefore, it is useful for larger areas. However, it is sometimes slower than the brush cutter and it is certainly more cumbersome.

The brush cutter is the best tool for thinning. It allows you to work more comfortably, more efficiently, and with less effort.

However, there is a price for that efficiency: it does require some training and good technique.

Work method

Work at right angles to the wind so that the trees fall properly.

Divide the area being thinned into imaginary strips 50 m (60 ft.) wide. Crisscross the entire width of the strip.

Work across an area 2 m (6.5 ft.) wide, advancing gradually.

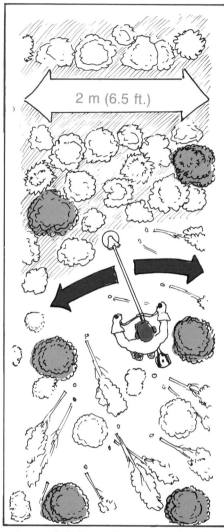

Ensure that the trees fall into the area that has already been thinned. Otherwise, you will quickly find yourself entangled in a mass of branches that impede your movement.

Watch out for sudden slopes or holes that can cause falls or skids.

Whichever tool you are using, be careful. Always wear the recommended safety equipment (helmet, visor, leggings, gloves, and so on). Always have a first-aid kit and a dry chemical extinguisher handy.

COMMERCIAL THINNING

What is it?

Commercial thinning is a cutting in stands in which the trees have a diameter at breast height greater than 10 cm (4 in.).

A stand can be thinned several times (at minimum intervals of 10 years) before the trees in it have reached maturity.

CHOOSE THE CROP TREES

Why thin?

Commercial thinning stimulates the growth of the released trees, increases their wood production, and makes up for natural mortality. It is necessary when the trees are affecting each other's growth.

You should release the best trees. With their competition removed, these trees will have more nutrients, light, and space.

The principle is the same as for precommercial thinning: select the crop trees and give them a helping hand.

When?

A stand should be thinned when the crowns of the trees are intertwined and no light is penetrating the stand.

The trees must be able to make use of more light, nutrients, or space.

You can judge whether it is time to thin by the size of the crown on the trees.

Thinning is not necessary when more than half the height of the trunk is covered in green branches.

It is too late to thin when less than one-fifth of the trunk is covered in green branches. There is no longer enough foliage to react.

The time to thin is when between half and one-fifth the trunk is covered in green branches and the branches of neighboring trees are intertwined.

Check the growth rings of the trees in the stand as well. If they have begun to shrink, it may be too late to thin; the stand will not react.

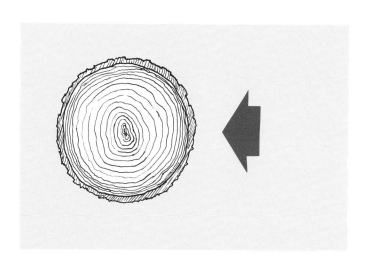

Choose the right tree

You must choose the right crop trees to release. Choose those that:

- belong to desirable species;

- have enough crown to respond to thinning;

- are not diseased (canker, rust) or otherwise defective (forks, broken branches); and

- have straight trunks.

How to thin

The aim of commercial thinning is to free the crop trees from competition.

Do not thin the stand too much as this may cause windfall, especially if it is the first thinning.

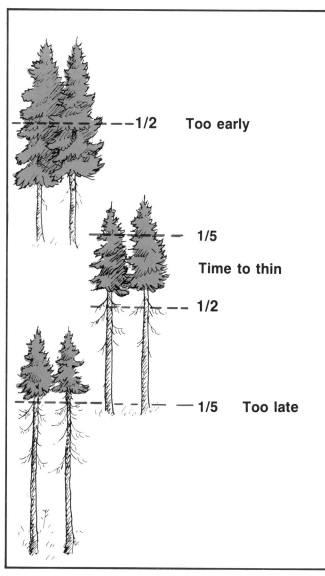

Release the crowns without leaving gaps.

To release crop trees, remove, in order of priority:

1°   *fallen, dead, dying, and diseased trees;*

2°   *mature trees and those that will be mature before the next thinning;*

3°   *trees that are too close to the crop trees.*

You can mark the trees that are to be removed. This will make the actual thinning easier. You will be able to concentrate on your work and reduce the possibility of error.

Identify the trees to be eliminated by marking the trunk and base with paint. After cutting, the paint on the stump will let you verify that the right trees were removed.

Work method

It is important to know what to cut; it is equally important to know how. You must provide skid trails for the removal of fallen trees.

Lay the skid trails in the direction of the slope.

The ideal distance between trails is 20 m (66 ft.).

By providing clearly marked trails, you eliminate excessive movement through the forest and reduce the damage to the crop trees that you so carefully selected.

The width of the trail will vary depending on the hauling equipment you use. Farm tractors, the F-4 Dion and J-5 require roads that are 3 m (10 ft.) wide.

When horses or snowmobiles are used, a road 1.5 m (5 ft.) wide is sufficient.

The trails should not be longer than 240 m (800 ft.). If they are longer than that, provide other secondary logging roads.

Work in an area within 10 m (33 ft.) of either side of a skid trail.

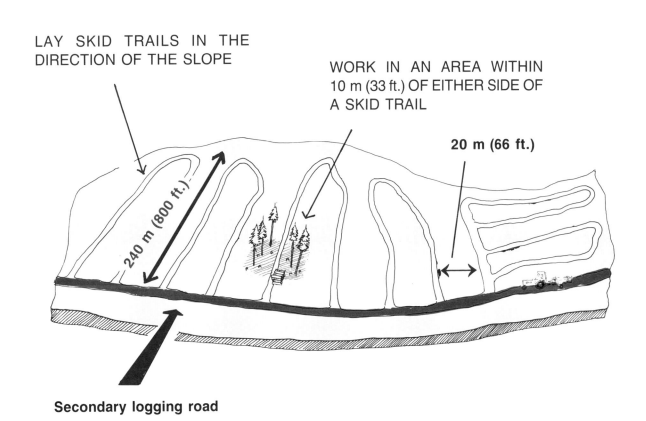

LAY SKID TRAILS IN THE DIRECTION OF THE SLOPE

WORK IN AN AREA WITHIN 10 m (33 ft.) OF EITHER SIDE OF A SKID TRAIL

20 m (66 ft.)

240 m (800 ft.)

Secondary logging road

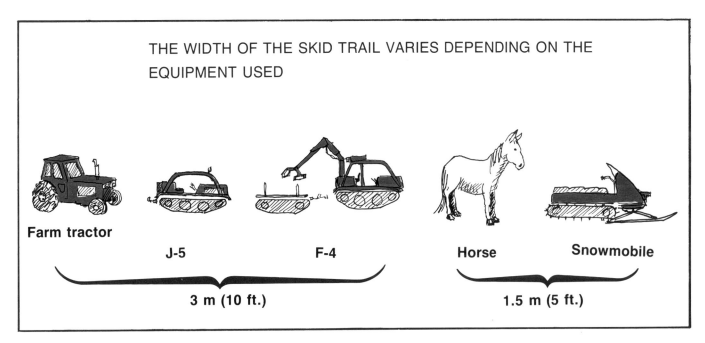

THE WIDTH OF THE SKID TRAIL VARIES DEPENDING ON THE EQUIPMENT USED

Farm tractor

J-5

F-4

Horse

Snowmobile

3 m (10 ft.)

1.5 m (5 ft.)

Keep skid trails free of all debris and cut the stumps close to the ground. This will prevent accidents and breakdowns. Remember that a tractor is not designed to do the work of a skidder.

However, on wet ground lay branches on the logging road; this will allow the machinery to move around more easily.

Finally, don't forget that trees cut to make skid trails are part of the thinning.

Work safely. Use well-maintained tools and appropriate safety equipment. Refer to the sections on work tools and methods and on wood harvesting.

Don't forget that the purpose of thinning is to improve your stands. It must be done properly because it affects the quality of the mature stand.

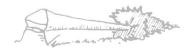

A well-planned and methodically executed timber harvest can facilitate a property owner's job and prevent accidents. Using proven work methods also ensures that the final products will be of good quality.

Timber harvesting

Working in the forest, you regularly encounter felling problems requiring techniques more involved than those covered in the section on «Work Tools and Methods». Often, you have to cut down a leaning tree or a snag that is so fragile the slightest breeze may knock it over. Other times, a felled tree may lodge in the crown of another or an obstacle close to the stump may present a potential hazard.

Proper work methods exist to help you deal with these problems. It is your responsibility to be familiar with them and to use them. Inappropriate techniques can result in a product of inferior quality or, worse yet, cause serious injury.

For greater safety, avoid working alone in your woodlot.

FELLING PROBLEMS

- *Tree leaning slightly in the direction of fall*

The problem:

When backcutting, acute tension is created in the holding wood (hinge). This tension can cause the trunk to break or splinter (barber's chair).

The solution:

Make an undercut in the normal manner (1/3 of the diameter with an upward angle of 45°). Then make a backcut in five sawcuts.

1. UNDERCUT
2. CUT MADE TO REDUCE TENSION IN THE HOLDING WOOD
3. BACKCUT N° 1
4. BACKCUT N° 2
5. FINAL BACKCUT

1/3 of the diameter

Maximum of 2 cm (1 in.)

Hinge (holding wood)

The problem:

There is a high risk that the stump will split.

The solution:

Make an undercut 1/4 of the diameter and then make a boring cut with the tip of the guide bar to shape the hinge (holding wood). Be prepared in case of kickback! Leave part of the trunk at the back to support the tree. Finally, make a backcut 4 cm (2 in.) lower than the boring cut.

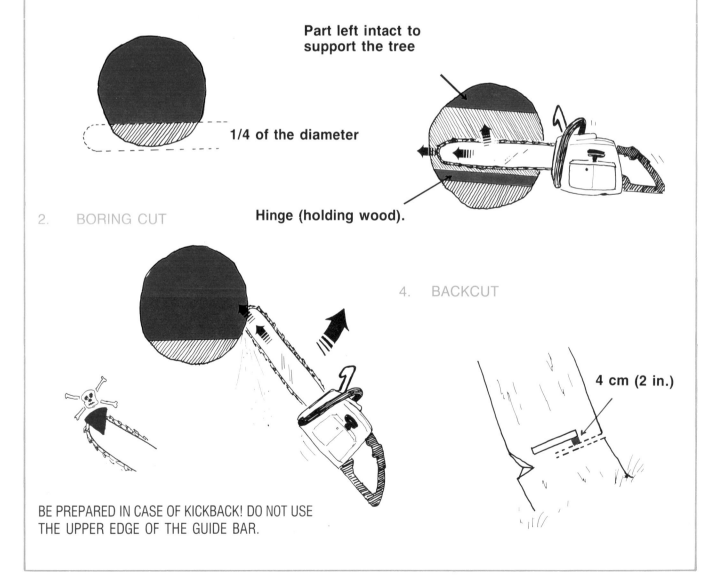

1.   UNDERCUT

3.   BORING CUT

Part left intact to support the tree

1/4 of the diameter

Hinge (holding wood).

2.   BORING CUT

4.   BACKCUT

4 cm (2 in.)

BE PREPARED IN CASE OF KICKBACK! DO NOT USE THE UPPER EDGE OF THE GUIDE BAR.

*   *Tree leaning in a direction other than the direction of fall*

The problem:

There is a chance that the tree will sit back on the guide bar when you make the backcut.

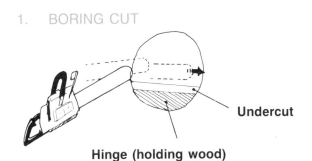

1. BORING CUT

Undercut

Hinge (holding wood)

2. INSERT THE WEDGES AND CONTINUE WITH THE BACKCUT

Wedge

The solution:

Use felling wedges and levers. First make the undercut, then make a boring cut leaving the hinge (holding wood). Once there is enough space, insert a wedge on each side of the tree. Finish the backcut by cutting toward the outside. Use the lever to force the tree over. The tree will fall more easily and with less effort.

3. FORCE THE TREE OVER

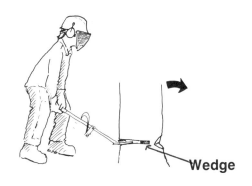

Wedge

*   *Tree whose direction of fall is uncertain*

The problem:

There is a chance that the tree will lodge in the crown of another.

The solution:

First remove the surrounding trees to reduce the chances of it becoming caught.

• *Lodged tree*

A lodged tree should be brought to the ground immediately. Never walk underneath the tree.

The solution:

Use a cant hook to pivot the tree or a lever to move it back. If you do not use one of these two methods, you could injure yourself.

Danger zone

Never attempt to bring a tree down by cutting the tree it is resting on or by felling another tree on top of it.

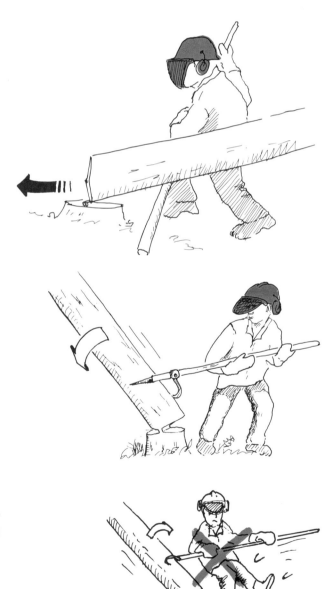

The problem:

These trees are very dangerous. Their dry branches can break and fall on a forestry worker.

The solution:

Ideally, snags should be pushed over with the back of a skidder or cut down with the help of another crew member, who can watch for falling branches.

A snag should never be left standing. Cut it down before other trees!

• *Obstacle close to the tree base*

The problem:

The undercut and backcut cannot be made in their usual spots.

The solution:

Clear away the surrounding area as much as possible and make the undercut and backcut higher than the obstacle.

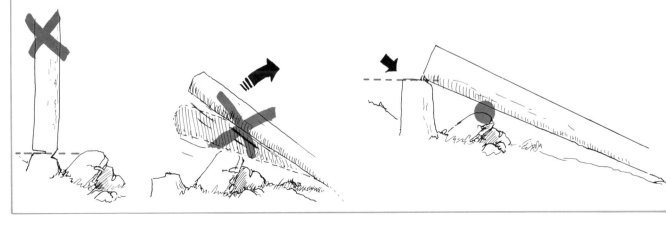

Planning the job

Now that you are familiar with the felling techniques, you must plan your forest operations to ensure they are carried out both safely and efficiently. The planning process should include:

- locating and marking the areas to be cut;

- walking on the site to familiarize yourself with the topography;

- identifying potential problem areas;

- locating and marking roads to remove the wood from the forest.

Planning is important for both the faller and the person in charge of removing wood. For the faller, planning is done tree by tree. For the hauler or skidder, it is done by group of trees.

Advantages of planning

- *Saving of time and energy and increased safety*

By planning the felling and using directional felling, you can reduce the amount of work required.

The direction of fall should be chosen so that the trees land on logs already on the ground or on small hillocks. This way, the fallen trees will be at a convenient height and limbing and bucking can be done in a more comfortable position.

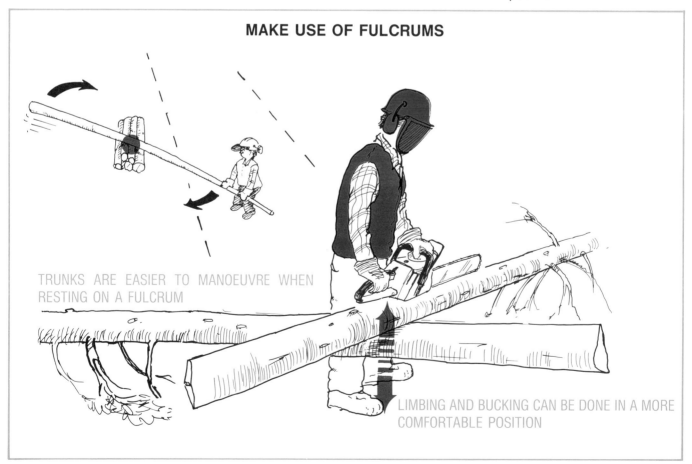

MAKE USE OF FULCRUMS

TRUNKS ARE EASIER TO MANOEUVRE WHEN RESTING ON A FULCRUM

LIMBING AND BUCKING CAN BE DONE IN A MORE COMFORTABLE POSITION

Another way of making your job easier is to cut down the larger trees close to stacks; this will make hauling distances shorter. The smaller trees can be felled close to piles of branches.

When no planning is carried out, 40% of the terrain can be scarred by machinery. This proportion should never exceed 15 or 20%.

Cut down the larger trees close to stacks

• *Soil protection and regeneration*

Machinery can cause damage to the soil and remaining trees, preventing the forest from regenerating.

Proper planning is needed to protect the soil and to facilitate regeneration of the forest. Damage is primarily caused by vehicles used to remove wood.

It is therefore important to plan roads properly. They must be as straight as possible so that vehicles can circulate easily.

In addition, the amount of traffic over watercourses should be limited. Areas where vehicles cross should be well indicated. Choose spots where the bed is rocky. Cross the watercourse at a right angle and always cross at the same spot.

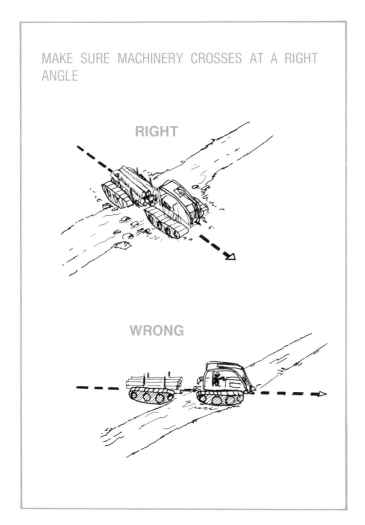

MAKE SURE MACHINERY CROSSES AT A RIGHT ANGLE

RIGHT

WRONG

Finally, try to control water-induced erosion of the wood roads by placing them along the contours of the terrain (perpendicular to the slope).

If the slope is too steep, however, the roads cannot be positioned this way since the danger of machinery toppling over is too great. In this case, the roads should follow the direction of the slope. Once the timber harvest has been completed, barriers can be formed to reduce the water velocity and thus limit erosion.

REDUCE THE RISKS OF EROSION ALONG WOOD ROADS BY FORMING BARRIERS

Timber harvesting

Now that you are familiar with the techniques required to cut down troublesome trees and have done your planning, it is time for action. But don't think your training is over yet! It is important for you to choose the harvesting method best suited to the kind of treatment you have in mind.

1) Clear away trees (2 or 3 wide) for the road. Place a few logs diagonally across the road without limbing or bucking them. Use them rather as fulcrums for limbing and bucking other trees.

2) Do this on one side of the road, then the other. The stacks will serve as fulcrums.

3) Place branches and logging (cutting) debris in the central area between two stacking areas.

4) Repeat these operations until you have covered the marked-off area.

1. USE THE FIRST FELLED TREES AS FULCRUMS

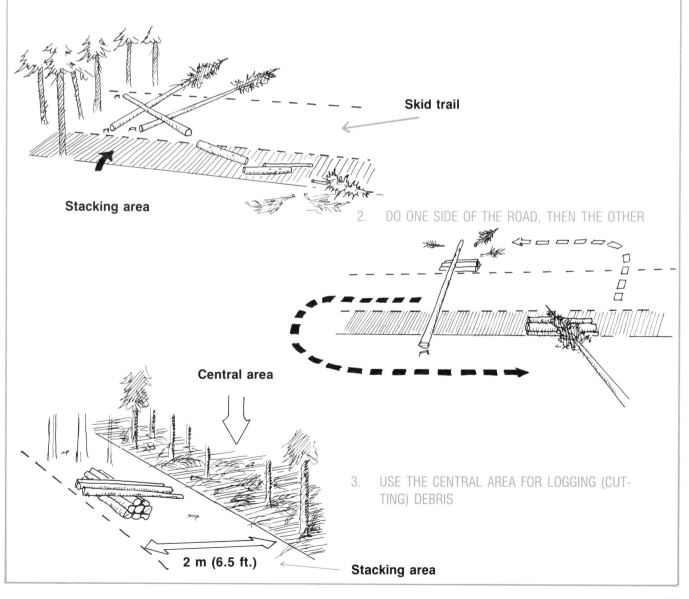

Skid trail

Stacking area

2. DO ONE SIDE OF THE ROAD, THEN THE OTHER

Central area

3. USE THE CENTRAL AREA FOR LOGGING (CUT-TING) DEBRIS

2 m (6.5 ft.)

Stacking area

1) Plan the cutting of one group of trees at a time.

2) Cut one tree in each of the stacking areas. Do not limb or buck them.

3) Cut the other trees using the first ones as fulcrums for limbing and bucking.

4) Continue this procedure until you reach the end of the road.

1. SELECT A GROUP OF TREES TO BE HARVESTED

2. CUT ONE TREE IN EACH OF THE STACKING AREAS

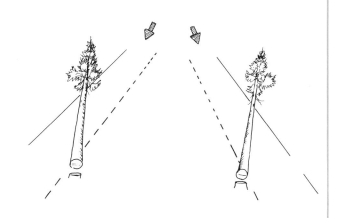

3. USE THE FIRST TREES CUT AS FULCRUMS

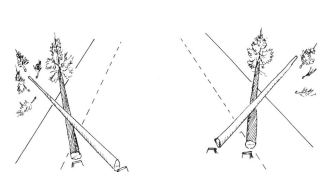

4. CONTINUE TO THE END OF THE ROAD

The difference between hauling and skidding

Hauling consists of transporting the timber wood using either a machine designed for this purpose or a trailer. This technique is used primarily for removing pulpwood and short saw logs.

Skidding is the process of sliding felled trees using grippers or winches. This method is used primarily for logs over 2.5 m (8 ft.) in length.

Removal of logs

Harvesting trees in private forests is primarily done using J-5 and F-4 Dion hauling machines or farm tractors with trailers. When these machines are used, the following distances should be respected:

- Distance between the roads: 20 m (66 ft.);
- Road width: 3 m (10 ft.);
- Width of stacking areas: 2 m (6.5 ft.).

The stacking areas are situated on each side of the road.

Free the roads of all logging (cutting) debris and cut the stumps close to the ground. If the terrain is wet, however, branches placed on the ground can improve its bearing capacity. Finally, when the terrain is steeply sloped, make the wood roads somewhat wider (4 m (13 ft.)). The skidder will cause less damage to residual trees (less likely to knock them over or split their bark).

If you use a skidder, the following distances should be respected:

- Distance between the roads: 20 m (66 ft.);
- Road width: 4 m (13 ft.).

In private forests, skidders are rarely used because they require a great deal of space to manoeuvre. They are never used for thinning or shelterwood cutting because wide roads are needed. Moreover, as it is extremely difficult to control the direction of skidding logs, they can cause damage to the residual trees and hinder regeneration.

On-the-job safety

You are now more familiar with the harvesting techniques and methods which will make your job easier and safer. The next item on the list is safety equipment. Do not forget that the equipment must be in good condition and that it is compulsory.

When felling trees, you should wear:

- *an approved safety helmet;*
- *a visor;*
- *ear protectors to reduce noise;*
- *safety boots;*
- *safety gloves or mittens;*
- *leg protectors (leggings or safety pants);*
- *dry chemical extinguisher;*
- *first-aid kit.*

Make sure that your chain saw is in good working condition and equipped with an emergency brake and anti-vibration handles.

Before filling up the fuel tank of your chain saw, make sure the motor has been turned off and you are not smoking.

Start up the motor only once you have moved away from the filling area and main fuel tank.

When removing wood from the forest

Machine operators must wear the same safety equipment as fallers. The gloves differ, however, in that they must be able to resist perforations caused by strands of steel cable.

Every precaution must be taken when using vehicles to remove wood from the forest. Avoid turning abruptly, especially when on a slope; otherwise the vehicle may topple over.

If you are working as a crew member, do not walk in an area where a machine is in operation. Be sure the brake is on when the vehicle is not in motion.

This section has presented the main techniques involved in timber harvesting. If you require further information, do not hesitate to contact your forestry adviser. You may also wish to obtain the brochures that have been produced on the subject by some chain-saw manufacturers. Ask your retailer.

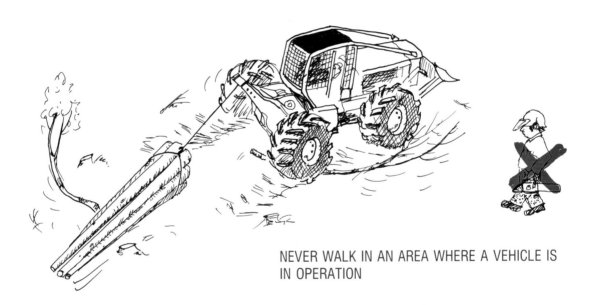

NEVER WALK IN AN AREA WHERE A VEHICLE IS IN OPERATION

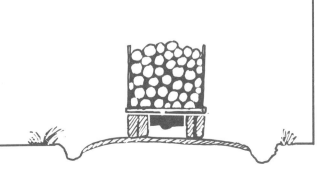

FOREST ROADS

The construction of a forest road is as much a part of woodland management as any other task. Your road should be both durable and easy to maintain. The following pages will show you how to go about this task.

FOREST ROADS

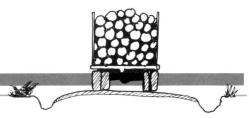

Building forest roads is not a silvicultural practice but rather a way to make your work easier.

A careful selection of the layout is important since the road will play a major role in forestry operations.

Furthermore, a forest road makes it easier to monitor and control fire, insects, diseases, and other woodland enemies and to move the equipment required for various management activities to the work site.

Each woodlot or group of woodlots should have a suitable forest road that allows access to every part of the lot.

Preliminary work

Some preliminary work is required before starting to build a forest road.

The first step is to determine the layout. As far as possible, the road should be built in the middle of the lot. If you own two adjoining lots, build the road along the line separating them.

After you have established the preliminary layout, study the land and identify any obstacles that will hinder the construction of the road (steep slopes, wet areas, rocky outcrops, and so on). Try to find the most practical solution.

Lastly, plan the final layout on the lot. Use line tape to mark the center and edges of the future road.

Forest road characteristics

The forest road should:

- allow access by a (ten-wheel) truck for the removal of wood;

- be within 120 m (400 ft.) of the front range line;

109

- have two turning areas, one L-shaped at the halfway point and one T-shaped at the end, where a truck loaded with wood can turn.

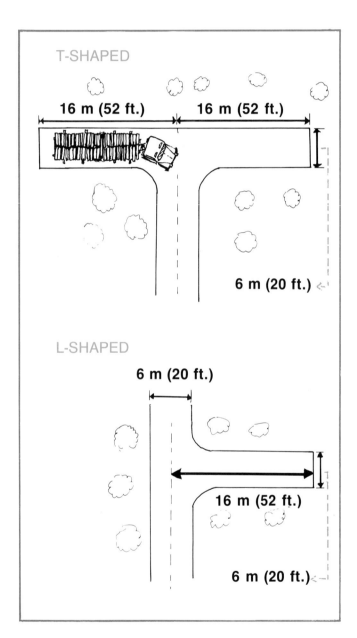

Road construction

The building of a forest road is a step that should not be neglected in the management of your lot. It is an expensive job that should be done properly.

The first step is to clear a 12-m (40-ft.) road allowance.

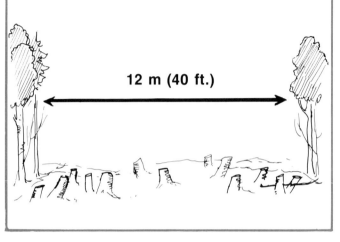

The road allowance should be wide enough for the road surface, ditches, and logging debris (slash).

Furthermore, this width will allow the sun to dry the road surface quickly. Remember, a road that remains wet will deteriorate more rapidly.

The road surface should measure 6 m (20 ft.) across, so that a truck may drive through when wood is stacked on the roadside.

A crawler tractor fitted with an angle blade should be used to build the road. The tractor removes logging debris (slash) and stumps and pushes them to either side of the road.

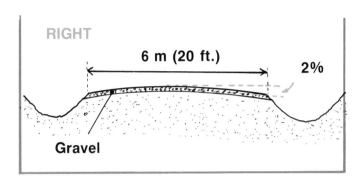

RIGHT

6 m (20 ft.)

2%

Gravel

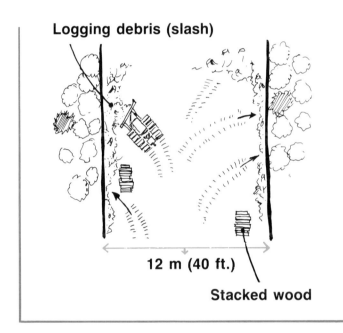

Logging debris (slash)

12 m (40 ft.)

Stacked wood

The road should then be leveled and shaped.

The road must be higher than the surrounding ground level. Avoid building sunken roads from which water cannot flow into the ditches.

AVOID

Lastly, the middle of the road surface should be higher than the sides or convex (slope of 2%) to allow the surface water to drain into the ditches.

To improve the road surface, cover it with 8 to 10 cm (3 to 4 in.) of gravel.

Road drainage

It is essential that water not stand on or flow onto the road surface.

Good drainage and a minimum of maintenance will help your road last longer.

Where the land is flat, a ditch is required on either side of the road. On sloping land, one ditch is sometimes sufficient. Remember to take into account the natural flow of water.

In general, the ditches should have the following dimensions:

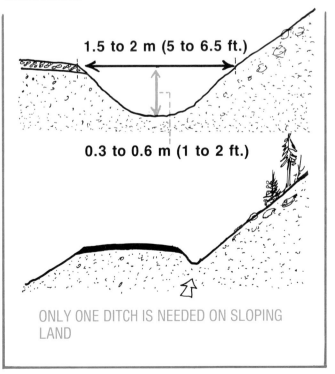

1.5 to 2 m (5 to 6.5 ft.)

0.3 to 0.6 m (1 to 2 ft.)

ONLY ONE DITCH IS NEEDED ON SLOPING LAND

Culverts are required when the road crosses an intermittent or perennial watercourse.

The culvert should be wide enough to allow for the maximum water flow. If the diameter is too small, the culvert will not be able to drain all the water and some of it will run onto the road surface and cause erosion.

Different types of culverts can be used according to the size of the watercourse.

PLASTIC CULVERTS

These are field drain pipes used for draining very small amounts of water.

Great care is required to ensure that they are installed in a straight line, with no bends.

Place logs on either side of the pipe to increase its strength and ensure uniform distribution of pressure. This will prevent the pipe from being crushed.

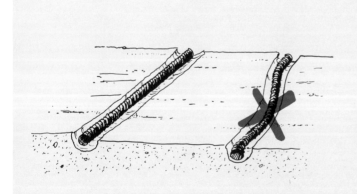
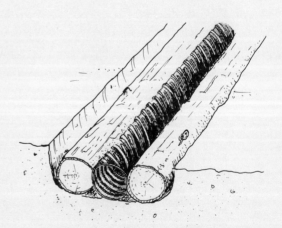

WOODEN CULVERTS

Wooden culverts are used when the water flow is slight.

They are made from two logs and have a deck of planks or logs.

CORRUGATED STEEL CULVERTS

The cylindrical shape of these culverts makes them very strong.

They come in several sizes and can be used when the flow of water is light or strong.

When properly installed, they are both strong and effective.

Position the bottom of the pipe a few centimetres (10 cm (4 in.)) below the bed of the watercourse.

Place fine, well-packed soil on either side of the pipe. This will ensure even distribution of the load over the entire circumference of the culvert.

Cover the pipe with a minimum of 30 cm (12 in.) of gravel.

30 cm (12 in.) of gravel

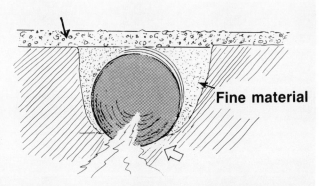

Fine material

10 cm (4 in.) below the bed of the watercourse

WOODEN BRIDGES

Wooden bridges are used to cross larger watercourses (streams, rivers, and so on).

They consist of two caissons and a wooden deck.

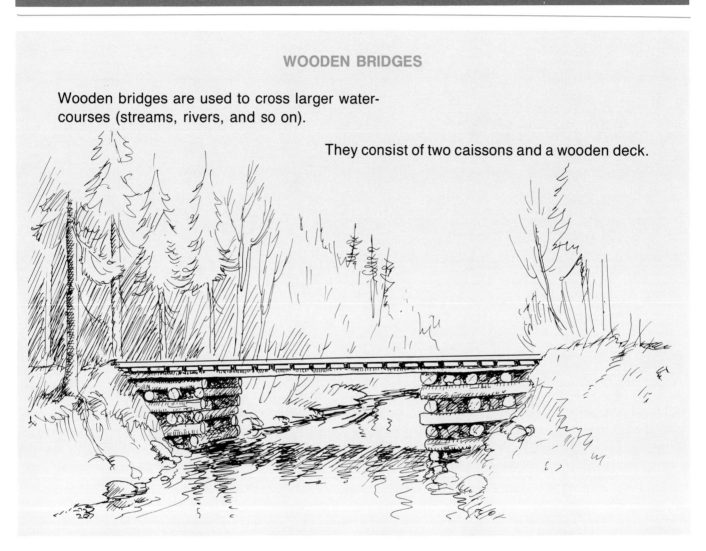

No matter what type of culvert is used, it should always be placed along the axis of the watercourse. Any change caused in the direction of the watercourse could lead to the erosion of embankments.

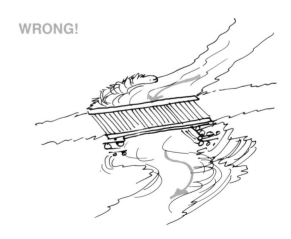

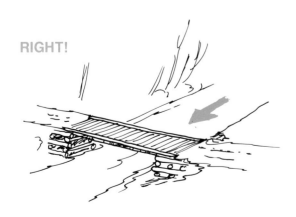

It is a good idea to build a stone wall at the openings of large culverts (those measuring 60 m (24 in.) or more).

114

Special cases

Wet Areas

Your lot may contain very wet areas. Try to skirt around them. If this is not possible, build your road where the ground appears most solid.

In small wet areas, place logs on the ground to make it more stable.

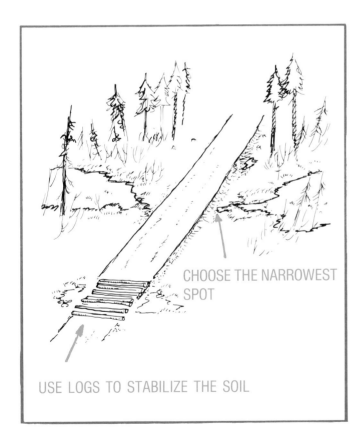

Use the mineral soil to cover the road surface. The excavated areas will serve as ditches. Allow this part of the road to dry out before covering it with gravel.

Sloping Land

Use the cut-and-fill method to build a road into the side of sloping terrain. This method will provide a ready supply of material and allow you to build a stable road.

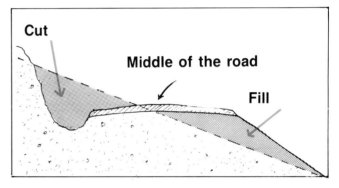

A ditch is needed on the uphill side only. Transverse culverts should be placed at regular intervals so that water can drain downhill.

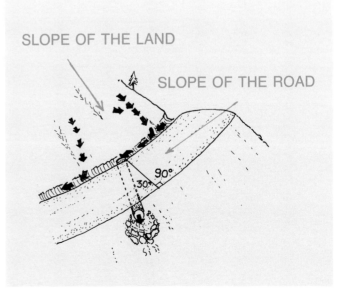

Improvement and maintenance

The most common road improvements are widening road surfaces and repairing culverts or ditches.

Maintenance should be done during the yearly road inspection. Clear any plant material and soil away from the ditches and the culvert openings.

Fill in holes and level bumps and ruts on the road surface. This can be done by dragging a blade behind a farm tractor.

Push the gravel from the road edges to the center so that the road surface remains convex.

By building a forest road, you will gain access to the various parts of your woodlot. A carefully built and regularly maintained road will last a long time.

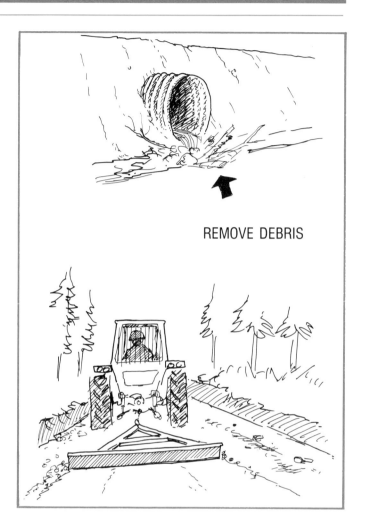

REMOVE DEBRIS

FOREST PROTECTION

Although a necessary operation, forest protection is often neglected. Anyone who owns a woodlot should take the time to carry out forest protection activities. It is important to keep insects and diseases from robbing you of the benefits of your woodlot.

FOREST PROTECTION

The purpose of forest protection is to counter the various forms of attack that a forest may be subjected to.

It is particularly intended to guard against insects, diseases, and fire.

However, we must remember that people can cause damage too, by improper cutting or by driving vehicles in woodlots. Allowing animals to graze in a woodlot can also be harmful. Some wild animals can cause damage as well.

All these types of damage decrease the value of your woodlot, both aesthetically and financially.

Forest protection comprises prevention and treatment.

Prevention

Damage caused by people can by limited mainly by taking preventive measures. Public access must be restricted, by posting notices or putting up fences, in order to reduce the chances of vandalism, especially in managed stands. If you or your neighbors have farm animals, put up fences to keep them from grazing in your woodlot. Also watch for damage caused by wild animals such as rabbits and field mice, and take steps if necessary (see section on plantation maintenance). Finally, ensure that there are roads, firebreaks (open areas without vegetation), and water supplies for fire-fighting.

There are also preventive measures to reduce the chance of attack by insects or diseases. Most plant diseases and insects are naturally present in your forest. Nevertheless, there are methods to reduce the damage they cause.

Think back to the sections on thinning and cutting. In each section, choosing which trees to fell and which ones to keep was discussed.

In all cases, diseased, malformed, or old trees were the ones to remove. Because they were weaker, they were more vulnerable to insects and diseases. You were also advised to leave companion species — that is, to maintain a mixed stand (for example, fir, spruce, and birch).

These silvicultural practices improve a stand's resistance to diseases and insects while encouraging tree growth.

By removing the weaker trees, there is less chance of diseases taking hold and spreading to healthy trees. In addition, maintaining several species in a stand makes it less attractive to insects because food is harder to find (insects are usually drawn to one type of tree in particular). Finally, healthy trees are more resistant to infestation and diseases.

Effective forest protection against diseases and insects depends on good silvicultural practices and harvesting methods.

Although forest protection decreases the possibility of diseases or infestation in your woodlot, it does not eliminate it entirely. Fortunately, treatment can save many trees and sometimes, in the case of epidemics, entire stands.

Treatment

In forest protection, treatment methods can be mechanical, chemical, or biological. They can eliminate a problem before it gets out of hand. Various types of treatment are described below.

Pruning affected limbs and trees

Pruning is a means of protection that consists of removing affected branches before damage spreads to the rest of the stand.

This technique must be used properly, since poor pruning can make the problem worse rather than solve it. Furthermore, it can slow down the growth of your trees. To determine the correct method to use, refer to the section on plantation maintenance.

If insects are attacking the leading shoot of your conifers, topping is the method to use. Remove the leader and the last whorl, leaving the best branch. This branch will replace the leader.

**TWO MAIN PRINCIPLES
OF FOREST PROTECTION**

1. KEEP THE FOREST MIXED

2. REMOVE THE WEAKER TREES

TOPPING A CONIFER

Leave the best branch

Cut here

Finally, another treatment method is sanitation cutting (see section on cutting). It is used when some trees are diseased or infested, and there is a danger of the condition spreading through the stand.

After pruning, topping, or cutting, it is sometimes necessary to destroy affected branches or trees to keep the disease or infestation from spreading. You can burn the branches or trees, or bury them. Be careful if you burn them. If possible, avoid dragging them through uncontaminated areas so that the infestation does not spread.

KEEP AWAY FROM UNCONTAMINATED AREAS

Use of chemicals or biological products

Use of chemicals and biological products is another method of combating insects and diseases. If your problem is caused by insects, use insecticides; if it is caused by fungi, use fungicides. Ask an expert to help you select the correct product.

Be careful when using these products. Use the product designed for the insect or disease that is causing the problem, and read the instructions on the label carefully. Apply the recommended amount — no more, no less.

Finally, use safety equipment such as waterproof clothing, boots, gloves, and masks.

Inspection

To do a good job of forest protection, you have to keep an eye on your woodlot, so inspect it regularly.

Inspection doesn't cost much, and it allows you to quickly detect anything in your stand that is out of the ordinary. Observe abnormal phenomena to see what develops. If you see that the problem is getting worse, consult your forestry adviser, who will be able to solve your problem before it gets out of hand.

The following table, showing the most common symptoms, will help you detect problems caused by insects or diseases.

There are illustrated guides as well that you can use to identify an insect or disease in your forest.

FOREST PROTECTION

CAUSE	PARTS OF TREE AFFECTED		
	FOLIAGE	TRUNK	OTHER PARTS
Insects	• leaves turned red • leaves curled up • leaves discolored • galls	• holes and tunnels • resin secretion	• sticky substance on foliage • foamy, viscous liquid (like spittle) • withered leading shoot
Diseases	• leaves discolored • leaves stained • leaves turned yellow or red • loss of leaves	• fungi • resin secretion • cankers • rot	• dead branches

Foamy liquid

Leaves curled up

Holes and tunnels

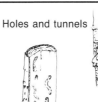

Canker

Resin secretion

Gall

Withered leading shoot

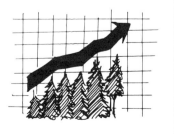

FINANCIAL MANAGEMENT OF A WOODLOT

The financial aspect of owning a woodlot is just as important as the forestry aspect. Yet it is all too often overlooked. As the owner of a woodlot, you owe it to yourself to manage your forestry operations wisely. You will discover how worthwhile such management can be. There are also certain acts and regulations regarding taxation and social benefits that may apply to your situation. It is important to know how they affect you.

FINANCIAL MANAGEMENT OF A WOODLOT

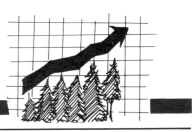

Forestry operations can sometimes entail major expenditures. It is useful to be able to assess the validity of these. Financial management involves comparing the amount of money earned from your woodlot with the amount spent on it. This is done for the purpose of obtaining a satisfactory return on your investment or increasing your profits.

FUNCTIONS OF FINANCIAL MANAGEMENT

Financial management has several functions, all of which are important to woodlot owners. It enables them to:

- monitor their financial operations;
- assess these operations;
- make plans; and
- simplify the preparation of reports.

The first step toward achieving these goals is adopting a basic accounting system geared to your forestry operations. There is nothing complicated about this and you will find it extremely useful.

Monitoring

With the help of a very simple accounting system, you will be able to keep up-to-date records of your revenues and expenditures. If you do this on a regular basis, you will always be able to quickly obtain a financial overview of your operations. By knowing at any given time what your activities cost and what they earn, you will be able to control them.

Assessing

For most property owners, a woodlot yields only a secondary income. You may wonder whether you should spend time and energy managing this part-time activity. You most certainly should. Sound financial management will enable you to learn the strengths and weaknesses of your forestry operations. It will then be easier to determine the types of measures you should be implementing.

Planning

For some property owners, financial management will be the start of genuine planning of their woodlot operations. Their accounting records will enable them to choose appropriate activities, not only from a forestry perspective but also from a financial perspective.

GOOD BOOKKEEPING WILL GIVE
YOU A BETTER PERSPECTIVE

Preparing documents

You occasionally have to prepare administrative documents. One of these is your income tax return, which must list income from your woodlot.

If you have kept a record of all receipts and expenditures, you will have no trouble assembling all the information required to complete your tax form.

In addition, good bookkeeping will provide you with convincing support when you apply for financing, such as a loan.

MAIN TYPES OF REVENUE AND EXPENSES

Revenue

The main sources of revenue are:

- Money from wood sales and cutting rights;

- Contributions and grants received by you from various management programs for your forestry operations;

- Receipts from the sale of forestry material, sugarbush activities, or rebates of municipal, school or excise gasoline taxes, and so on; and

- Amounts held, as applicable, for the working capital fund of the wood producers' associations.

Expenses

Forestry operations entail such expenses as:

- Equipment maintenance and operation (parts, gas, oil, repairs, insurance, registration, and so on);

- Employees' wages;

- Deductions at the source from emplyees' gross earnings (federal and provincial taxes and employee contributions to the government pension plan and Unemployment Insurance) and fringe benefits (employer contributions to the government pension plan, health insurance plan, Unemployment Insurance, the appropriate provincial agencies of health and safety, and also the provincial labor boards or commissions);

- Rental of machinery, buildings, and sites;

- Improvement and construction of forest roads and culverts;

- Payment to a contractor for carrying out work;

- Purchase of herbicides, silvicides, and other chemical or biological products;

- Purchase of equipment and supplies (gloves, boots, waterproof clothing, helmets, and other safety gear); and

- Other expenses, such as telephone bills, travel cost, interest on a loan, and so on.

If you are a farmer and you use your farm tractor for forest operations, your woodlot accounting should include only those expenses related to forest work. For instance, if 1/3 of your tractor's use is for forest operations, you should list only 1/3 of its total maintenance and operation costs.

CLEARLY IDENTIFY THE EXPENSES RELATED TO YOUR FOREST OPERATIONS

SUPPORTING DOCUMENTS

You should retain proof of all revenue and expenses. Remember that expenses with no supporting documents may be disallowed during a tax audit.

For every expense, you should therefore have a corresponding invoice and proof of payment. The best proof is your canceled check. Moreover, it is a good idea to maintain a separate checking account for financial operations related to your woodlot.

All revenue must be supported by an invoice for the sale of wood, a receipt, or a grant notice.

But, other than for preparing tax returns, why is it important to accurately determine your revenue and expenses? How can it help you?

PREPARING A STATEMENT OF REVENUE AND EXPENSE

Every year, you should prepare a statement of your revenue and expenses. This record will give you a year-end overview of how you earned your money and how you spent it. The final figures will show you whether you have made a profit or sustained a loss. That alone is reason enough to assess your revenue and expenses.

Over the years, you are bound to show losses in some fiscal periods. But that is no reason to give up your work. Forests take years to grow and require patience. Improvement expenses often yield returns only years later.

And, as shown in the following pages, it is possible to recover some of these losses.

A FEW FACTS

These items may be of interest to you as a woodlot owner. Some of them could mean significant tax benefits for you.

Fixed assets and depreciation

The purchase of fixed assets (durable goods) should not be treated as an expense at the time they are acquired. Instead, the cost of fixed assets should be deferred over several years. This distribution is called depreciation and should be calculated annually according to the type of good.

Depreciation thus represents a good's purchase cost averaged over the duration of its useful life. For example, the depreciation on a chain saw purchased for $700 can be calculated at the rate of 30 percent a year.

Given the complexity of these calculations, you may have to consult an accountant.

Capital gains

A capital gain usually results from the sale of something that has been owned for a long time. For example, the sale of farm buildings or the sale of land including the principal residence can be considered capital gains.

In some cases, the granting of cutting rights may be considered a capital gain. In certain circumstances, income from clear cutting by an outsider on your lot can be considered a capital gain, since it could take several decades after the felling for the site to become productive again.

We strongly recommend that you consult an expert to help you ascertain whether the sale of wood or other goods constitutes a capital gain.

If you do show a capital gain, remember that you can claim a tax exemption for it.

DO NOT HESITATE TO CONSULT AN ACCOUNTANT

Business losses

A business loss occurs when your expenses exceed your revenue for the year.

In such cases, you may take advantage of special provisions. On your tax return, the loss may be deducted from the total of your other income, whether from employment, investment, or sales.

Consult an accountant to ascertain whether any of the above items apply to your situation and, if they do, to help you make maximum use of them.

With these basic concepts, you can now consider implementing financial management.

ABOUT TAXATION

All individuals who earn income from forestry operations on their woodlot have certain tax obligations, as explained below.

Self-employment

Property owners who work their woodlots themselves are referred to as **self-employed.**

They are not hired by employers, but work for themselves. Therefore, the income from their forestry operations is not subject to any deductions at the source. In other words, there are no deductions from their earnings for federal or provincial tax, the government pension plan or Unemployment Insurance.

Farmers and contractors may also be considered self-employed.

Business income and taxes

According to the Act, income from self-employment, whether it is income from sales of forest yields, cutting rights or grants, is considered **business income**. It is included with your other income and must be listed on your tax return. However, expenses incurred in generating this income may be claimed as deductions.

Remember that, if your expenses exceed your income, you may take advantage of special tax provisions.

Following is some information on legislation and government programs that may be of use to you.

LEGISLATION AND GOVERNMENT PROGRAMS

Unemployment Insurance Act

All unemployment insurance claimants must declare their net earnings from their woodlot operations. This can be done on the "Claimant's Report Card". Net earnings are income from forestry operations less the expenses required to obtain them.

Government grants for silvicultural work cannot be considered employment income. This financial assistance thus has no remunerative value for the purpose of unemployment insurance benefits.

If you work part-time on your lot and remain available for work, you may still be eligible for unemployment insurance benefits.

For more information, contact the Canada Employment Centre closest to you.

Social Aid Act

Generally, social assistance recipients must advise the provincial department responsible for social assistance of the starting date of the work on their lot, the predicted duration of the work (the claimant's activity period), and revenue received from their woodlot.

For more information, contact your provincial department responsible for social assistance.

Health, safety, and job security

A private woodlot owner may obtain protection against work-related accidents or illness by contributing to a provincial agency responsible for health and safety on the job and job security. Also, in case of an accident or illness, you may be entitled to compensation. This is well worth looking into.

The nature of this document does not allow us to give a list of all the services available in each Canadian province. For more detailed information, contact your appropriate department or agencies.

CONVERSION TABLE

LENGTH	**1 mi.:**	1.6 km 1 609 m	**1 km:**	0.6 mi 3 281 ft.
	1 arp.:	58.5 m	**1 m:**	3.28 ft.
	1 ft.:	0.305 m		

AREA	**1 ac.:**	0.404 ha 4 047 m²	**1 ha:**	2.92 arp.² 2.47 ac.
	1 arp².:	0.342 ha 3 419 m²	**1 m²:**	10.8 ft.²
	1 ft.²:	0 093 m²		

VOLUME

SOLID: **(TRUE)**	**1 cu.:** 2.83 m³	**STACKED:** **(BULK)**	**1 cd.:** 3.62 m³
	1 m³: 0.35 cu.		**1 m³:** 0.275 cd.

ABBREVIATIONS

acre:	ac.	hectare:	ha	cubic metre:	m³
arpent:	arp.	kilometre:	km	mile:	mi.
cord:	cd.	metre:	m	foot:	ft.
cunit:	cu.	square metre:	m²		

EXAMPLES:

1 mi.   =   1.6 km
×
÷

1 ha   =   2.47 ac.
×
÷

Where: **6 mi.** = 6 × 1.6 = 9.6 km

4 km = 4 ÷ 1.6 = **2.5 mi.**

3 ha = 3 × 2.47 = 7.41 ac.

25 ac. = 25 ÷ 2.47 = **10.1 ha**

NOTE: Some factors are rounded or approximate.